Chemistry in Balance

'O' Level Edition

Chemistry in Balance

'O' Level Edition

A M Hughes
Head of Chemistry, Tavistock School

Adapted for 'O' Level by

J C Sadler
Chief Examiner for 'O' Level Chemistry, University of Cambridge Local Examinations Syndicate. Chairman, National Working Party for 16+ Chemistry.

F Burke
Subject Officer for Chemistry, University of Cambridge Local Examinations Syndicate.

University Tutorial Press

Published by University Tutorial Press Ltd
842 Yeovil Road, Slough, SL1 4JQ

Published 1984
Reprinted 1984
© A. M. Hughes, J. C. Sadler, F. Burke 1984

ISBN 0 7231 0854 4

Typeset at The Pitman Press, Bath, England

Printed by Spottiswoode Ballantyne Ltd, Colchester & London

Foreword

In recent years, a considerable number of chemistry texts have appeared which provide good resource material for teachers but often ignore the needs of students. 'Chemistry in Balance 'O' Level Edition' sets out to provide a concise, yet thorough treatment of material to reinforce the teaching programme. It must always be remembered that a good book is no substitute for a good teacher, but it can be an invaluable aid.

The syllabuses of most examination boards have undergone substantial modification in recent years and this edition aims to reflect those changes and also take into account those proposals made by the National Working Party on 16+ chemistry.

Most chapters contain details of experiments which give an indication of the range of practical work that could be carried out. Teachers will, no doubt, supplement those in the text with further experiments that they have devised themselves or have used successfully over a number of years. Each chapter concludes with a variety of questions that are designed to test knowledge, understanding and application and these play an important role in the learning process.

The ASE is shortly to publish an updated version of 'Chemical Nomenclature, Symbols and Terminology (1979)'. The proposals in that publication have been taken into account in preparing this edition and the naming of compounds will, therefore, be consistent with future examination papers.

Contents

Introduction

Ever since the first caveman burnt wood to keep warm, chemistry has affected our lives. Nowadays, chemistry is involved in making our food, our medicines and our clothes. Without the work of chemists there would be no radios, televisions, computers, satellites, detergents, plastics, pesticides, bombs or missiles. Chemistry is a very important subject. Its influence on our life is almost certainly going to increase. We cannot ignore it.

This book aims to introduce you to chemistry as simply as possible. Each chapter deals with an important topic in chemistry. By working through these topics you should get some idea of the way in which chemists think. You should realise some of their achievements and recognise some of their problems.

We hope that you will enjoy working with this book and that you find it helpful for your school chemistry course.

We would like to thank the following Examination Boards for permission to include past examination questions:

University of London School Examinations Department [L]

The Joint Matriculation Board [JMB]

University of Cambridge Local Examinations Syndicate [C]

Safety statement

It is not intended that pupils should attempt any of the experiments described in this book, without first obtaining permission and any necessary guidance from their teacher. It is assumed that any experiment with a potential hazard (eg burning hydrogen in air) or using potentially hazardous chemicals (eg sodium, potassium, concentrated acids etc.) will be demonstrated by the teacher.

Teachers may find the following publications useful:

1. *Hazards in the chemical laboratory* edited by G.D. Muir published by The Chemical Society.

2. *Hazardous Chemicals: A manual for Schools and Colleges* by Scottish Schools Science Equipment Research Centre published by Oliver and Boyd.

3. *Cleapse HAZCARDS* published by CLEAPSE development group.

4. *Safety in Science Laboratories* Department of Education and Science.

1 What are things made of?

Think of hard substances like diamond and steel. Think of a soft substance like foam rubber.

Think of liquids that pour easily, like water, and liquids that hardly pour at all, like treacle. Imagine what would happen if you hit each of these with a hammer. They would behave very differently.

There are so many different things with so many different properties that it is hardly surprising that we ask, "What are things made of?" This has been asked for thousands of years. Many suggestions have been made to answer it. Gradually scientists have come to believe that all things are made up of tiny particles. No scientist has ever seen one of these particles, they are far too small for that, but many simple experiments suggest that matter *must* be made up of particles.

1.1 Some evidence for particles

There are some simple experiments that make scientists think all things are made of particles.

Mixing nitrogen dioxide and air

Nitrogen dioxide is a brown gas and so it can clearly be seen. It is denser than air. If the apparatus in Fig 1.1 was left for about 15 minutes, you would find that the nitrogen dioxide and the air mix completely in both experiment 1 and experiment 2. This is shown in Fig. 1.2.

It may seem difficult to explain this unexpected result. However, if the gases are made up of *tiny particles, moving in all directions,* they will mix. The fact that nitrogen dioxide particles are *heavier* than air particles does not stop them mixing.

1.2 Solids, liquids and gases

Matter is almost certainly made up of tiny particles, but matter is peculiar stuff. Many substances can be solids, liquids or gases at different temperatures. Solids, liquids and gases are known as the three *states of matter.*

These states have some very different properties. Think of water as an example. Ice is hard and brittle and cannot be compressed. Water flows and has a shape which depends on the shape of its container. It cannot be compressed much. Steam, like water, has no shape. It completely fills any container and is easily compressed.

These differences can be explained in terms of particles.

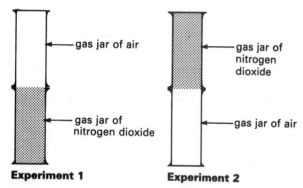

Experiment 1 — Experiment 2

Fig 1.1

gas jar of air

gas jar of nitrogen dioxide

gas jar of nitrogen dioxide

gas jar of air

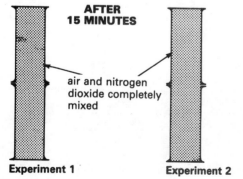

AFTER 15 MINUTES

air and nitrogen dioxide completely mixed

Experiment 1 — Experiment 2

Fig 1.2

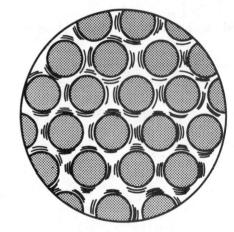

Fig 1.3 Solid

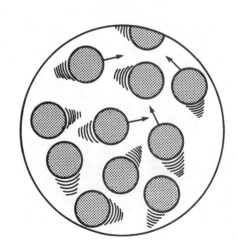

Fig 1.4 Liquid

In any solid the particles are close together. In crystals they are packed in a regular pattern. The particles are anchored around fixed positions. They can only vibrate about these fixed positions. The vibrations become stronger as the temperature increases. Particles in a solid are strongly attracted to each other.

In liquids the particles are close together. They are free to move within the liquid, but are attracted to the other liquid particles. The particles move faster as the temperature increases.

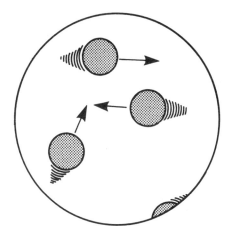

Fig 1.5 Gas

In gases the particles are usually far apart. They are free to move. The particles move so fast that there is little attraction between particles. The particles travel faster as the temperature increases.

This idea about the way in which particles behave in solids, liquids and gases is known as the *Kinetic Theory*. It describes all matter as being made up of particles in motion.

Table 1 shows the important differences between solids, liquids and gases. You should now be able to understand how these differences in properties can be explained by using the idea of particles.

1.3 Changes of state

Whether a substance is a solid, a liquid or a gas depends on temperature. We say water is a liquid because that is how it exists at room temperature. However, every day in kitchens water is changed from one state to another. Can you think where and when in your kitchen these changes take place?

We give names to these changes of state. These names are shown in Table 2.

Change of state			Name of change
Solid	$\longrightarrow$	Liquid	Melting
Solid	$\longrightarrow$	Gas	Sublimation
Liquid	$\longrightarrow$	Gas	Evaporation
Liquid	$\longrightarrow$	Solid	Freezing
Gas	$\longrightarrow$	Liquid	Condensation
Gas	$\longrightarrow$	Solid	Sublimation

Table 2

Let us look at the tiny particles of a substance when these changes take place (see Fig 1.6 and Fig 1.7 on page 4).

Melting/Freezing

When a solid is heated it melts. The particles gain energy and vibrate so strongly that they break away and become liquid particles. The solid becomes smaller and smaller and more and more liquid is formed.

When a liquid freezes, the particles move so slowly that they stick together when they collide. Solid crystals are formed. More and more particles stick onto the crystals until all the liquid has frozen.

Evaporation/Condensation

In the liquid state the particles move around but are attracted to other particles. Some have enough energy to break away and become a gas.

In the gas state the particles are free to move. When the particles collide with a liquid they are attracted to them. Some bounce off, others form more liquid.

Evaporation takes place when particles leave the liquid state *faster* than they join it from the gaseous state. A muddy puddle dries up because once water particles have found enough energy to escape from the liquid state they are blown away. They are unlikely to be captured by the water in that puddle again.

Condensation takes place when particles leave the liquid state *slower* than they join it from the gas state.

Solids	Liquids	Gases
Have high density	Have high density	Have low density
Cannot be compressed	Cannot be compressed	Easily compressed
Have fixed shape	No fixed shape	No fixed shape

Table 1

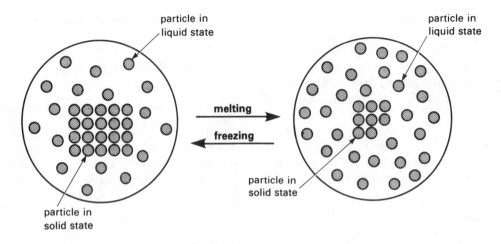

Fig 1.6 Melting and freezing

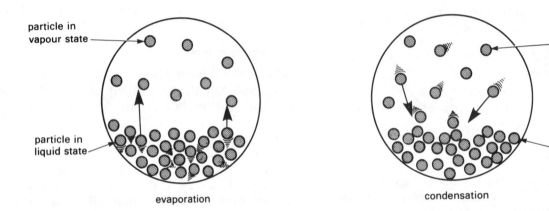

Fig 1.7 Evaporation and condensation

Sublimation

A solid can change directly to a gas, without going through the liquid state. This means particles break off the solid with enough energy to exist as gas particles (see below). This is not as unusual as you might expect. Think of any solid that smells. Particles must escape from the solid and become gas particles or you could not smell them.

A gas particle can also slow down enough to become a solid particle without going through the liquid state. This is why freezer compartments of fridges 'ice up'. Water vapour particles collide with the very cold surface and change directly into solid particles.

1.4 Dissolving

Where does the sugar go when you stir it into a cup of tea? The whole drink becomes sweet so the sugar must spread throughout the liquid. Is this more evidence that matter is made up of small particles? We can easily explain the process of dissolving using the idea of tiny particles.

Particles of the liquid (*solvent*) collide with particles of the substance being dissolved (*solute*). When they collide, they attract each other. Solvent particles pull off solute particles from the solid solute. The solvent particles surround the solute particles. This stops them rejoining. As the solvent particles move, the solute particles spread through the solution.

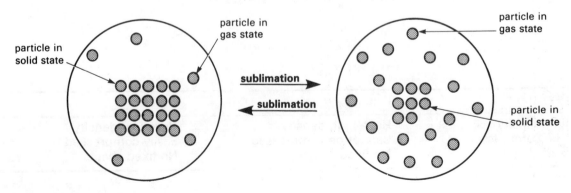

Fig 1.8 Sublimation

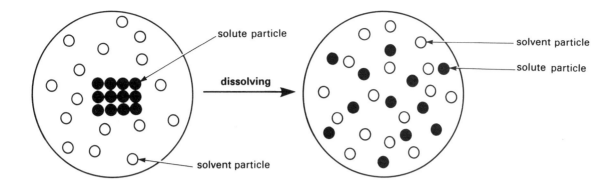

Fig 1.9 Dissolving

1.5 Dilution

Although it is not possible to measure the size of particles accurately, the following experiment will give you some idea about the size of particles.

Experiment 1.1 To show that particles of a solute are small

Dissolve 1 g of potassium manganate(VII) in 1 dm³ of water. Add one drop of this solution to 200 cm³ of water. Is the solution still coloured?

If there are 20 drops of potassium manganate(VII) solution in 1 cm³ of solution, work out the number of grams of potassium manganate(VII) in 200 cm³ of the diluted solution.

We can explain these observations if potassium manganate(VII) consists of very small particles. As the solution is diluted, the particles spread out throughout the whole of the volume.

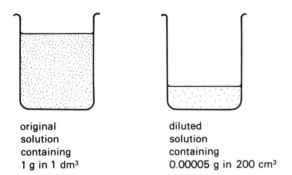

original
solution
containing
1 g in 1 dm³

diluted
solution
containing
0.00005 g in 200 cm³

Fig 1.10

1.6 Diffusion

When two substances mix due to the random movement of their particles, the process is called *diffusion*. We have already said diffusion is difficult to explain unless the idea of small particles is used. Fig 1.11 shows how we believe diffusion takes place.

Gases diffuse faster than liquids because there are greater spaces between the particles in gases. Mixing is therefore easier.

We should be grateful that gases diffuse quickly:
1. Unpleasant smells quickly disappear.
2. Gas leaks at home or school can often be quickly detected before any dangerous build up takes place.
3. If no diffusion took place in the Earth's atmosphere, the most dense gas might sink to ground level. The oxygen layer might well be out of our reach.

Diffusion and density

The speed at which the particles of a gas move depends on the mass of the particles. The lighter the particles the faster they move at any fixed temperature. Since hydrogen gas has the lightest particles they must be the fastest moving. This means that hydrogen will diffuse or mix faster than any other gas. You can show this by filling balloons; one with hydrogen and another with air. If these balloons are left, the one containing hydrogen will deflate faster than the one contain-

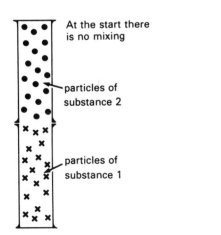

At the start there is no mixing

particles of substance 2

particles of substance 1

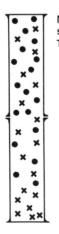

Moving particles are starting to mix. The substances are diffusing.

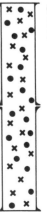

Complete mixing. Diffusion is complete, but the particles are still moving.

Fig 1.11

ing air. Hydrogen particles diffuse out of the balloon faster than any of the particles in air.

The problem with this simple experiment is that you have to wait a long time for the result. You could show that hydrogen diffuses faster than air more easily using the apparatus shown in Fig 1.12.

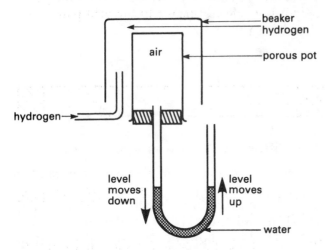

Fig 1.12 Diffusion of hydrogen

When the apparatus in Fig 1.12 is set up, some hydrogen molecules move into the porous pot and some air particles move out. Since hydrogen diffuses in faster than air diffuses out the number of particles in the porous pot increases. Therefore the pressure of gas in the porous pot increases. The water in the U-tube moves because of this increase in pressure.

Can you suggest what would happen to the water level in the U-tube if the hydrogen were replaced by a gas denser than air, such as carbon dioxide?

1.7 Are particles the answer?

On page 2 we asked "What are things made of?". Tiny, moving particles is an answer, but certainly not the full one. When scientists decided that matter was made up of tiny particles they could explain a number of observations, like melting and dissolving, which could not be explained before.

But many other questions then had to be asked: Questions like, "How big are the particles?", "How heavy are they?", "Why do they attract each other?", "How can particles of one substance be changed into particles of another substance?".

Partly by wanting to know the answer to these questions and by wanting to make better use of the Earth's resources, the science of chemistry has developed.

Questions

1. A boiling tube containing a colourless liquid **L** was placed in a beaker of boiling water. The liquid **L** started to boil almost immediately.

The boiling point of the liquid **L** is:

A less than 0 °C
B between 0 °C and room temperature
C between room temperature and 100 °C
D 100 °C
E above 100 °C.

2. Condensation occurs when a:

A liquid turns into a solid
B solid turns into a liquid
C solid turns into a vapour
D vapour turns into a liquid
E vapour turns into a solid.

3.

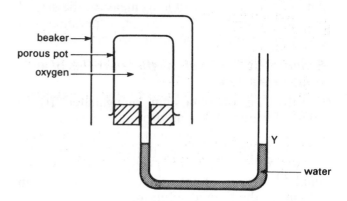

Which one of the following gases, when present in the beaker over the porous pot, will cause the water level at Y to fall?

A helium, He
B hydrogen, H_2
C carbon dioxide, CO_2
D methane, CH_4
E nitrogen, N_2

4. Nitrogen dioxide is a brown gas, and is denser than air.

A gas jar containing nitrogen dioxide is inverted on top of a gas jar containing air. Which one of the following correctly describes the colours inside the gas jars after a long period of time?

	Top	*Bottom*
A	colourless	dark brown
B	brown	brown
C	dark brown	colourless
D	dark brown	light brown
E	light brown	dark brown

5. Which one of the following provides the best evidence for the particle theory of matter?

A chromatography
B dehydration
C diffusion
D neutralisation
E oxidation

6. BOILING, CONDENSATION, DISSOLVING, DIFFUSION, EVAPORATION, FREEZING, MELTING, SUBLIMATION.

Choose from the list, the word which best describes the process taking place in each of the following:

For example Clothes drying on a clothes line—evaporation.

a Water forming on a kitchen window while food is being cooked

b Water changing to ice

c Sugar is stirred into a cup of tea

d Molten metal solidifies in a mould

e The smell of a 'stink bomb' gradually disappears

f A small puddle gradually dries up in warm weather

g Water changes to steam at 100 °C

h Iodine changes from a solid to a gas without becoming a liquid

i Ice forms from water vapour on the freezer compartment of a fridge

j Bubbles of ethanol vapour form in liquid ethanol

7. A solid compound X was heated steadily for 20 minutes. Its temperature varied as shown in the graph

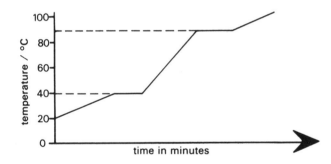

a At what temperature did X melt?

b What is the boiling point of X?

c What is the highest temperature at which X can exist as a solid?

d Would you expect X to be a solid, a liquid, or a gas at the following temperatures: (i) 25 °C, (ii) 50 °C, (iii) 100 °C?

e At what temperature does X exist as a solid *and* as a liquid?

8. Explain the following in terms of particles:

a Gases are easily compressed

b Solids need large forces in order to bend or break them

c 1 dm^3 of ice has a mass of approximately 1000 g whereas 1 dm^3 of water vapour has a mass of about 0·75 g

9. What do you understand by "*diffusion*"? Suggest a reason why, under similar conditions, carbon monoxide, nitrogen and ethene all diffuse at the same rate.

Explain the following observation.

A plug of cotton wool was soaked in concentrated ammonia solution and then placed in one end of a long glass tube and, at the same time, a similar plug soaked in concentrated hydrochloric acid was placed at the other end. A white deposit slowly formed in the tube in the position shown (nearer to the acid plug).

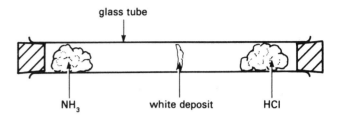

(Relative atomic masses: H, 1.0; N, 14.0; Cl, 35.5) [C]

10. The following experiment was carried out to illustrate diffusion. A few drops of concentrated nitric acid were added to pieces of copper in the bottom of a tall gas jar. A red-brown gas was formed and all the copper reacted in a very short time. The red-brown colour slowly moved up the gas jar but initially the colour remained darkest at the bottom. After about 30 minutes, the red-brown colour was uniform throughout the gas jar.

a Explain the meaning of *diffusion*

b Name the red-brown gas

c Why is the gas initially darkest at the bottom?

d Explain how and why the results would differ if:
   (i) the gas jar were placed in hot water before carrying out the experiment,
   (ii) bromine had been used instead of copper and concentrated nitric acid.

e Bromine vapour can be easily liquefied by cooling the vapour to 0 °C. Explain what happens to the bromine molecules as the vapour liquefies.

(Relative atomic masses: N, 14; O, 16; Br, 80)

2 Elements, Compounds and Mixtures

Substances are made up of very small particles called *atoms*. Although there are millions and millions of different substances in the world, there are only just over a hundred different sorts of atom.

Elements

An element is a substance that contains only one type of atom. The diagrams below represent: (i) an element in the gas state; (ii) a metal element in the solid state.

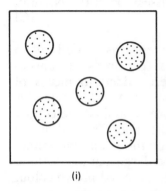 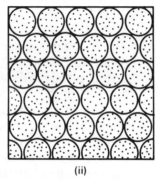

Fig 2.1

In some elements, atoms are joined together in clusters.

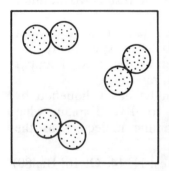 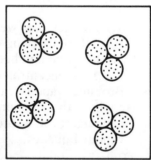

Fig 2.2

These clusters of atoms are called *molecules*.

Compounds

A compound is *always* made up of more than one type of atom. These different atoms are joined together in clusters. The diagrams in Fig 2.3 represent two different compounds existing as molecules.

In compound 1, one of the 'dark' coloured atoms is always joined to just one of the 'light' coloured atoms. So, in any sample of this compound there will always be equal numbers of 'light' and 'dark' atoms.

In compound 2, one of the 'dark' coloured atoms is always joined to three of the 'light' coloured atoms. So, in any sample of this compound, there will always be three times as many 'light' atoms as 'dark' atoms.

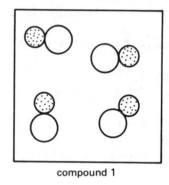 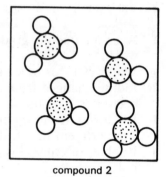

compound 1   compound 2

Fig 2.3

Mixtures

In a mixture of two elements there are two types of atom present but they are not joined to one another.

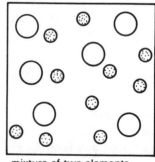 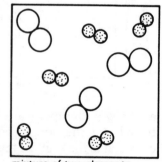

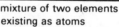

mixture of two elements existing as atoms   mixture of two elements existing as molecules

Fig 2.4

A compound of two elements always contains the same proportion of each element—you cannot alter it. However, a mixture can have any proportion of each element.

Alloys are mixtures of metals and other elements that have been melted together and then allowed to solidify. Some alloys are described on page 79. The diagrams in Fig 2.5 represent the structures of two solid alloys.

The structure on the left shows an alloy that is mainly a metal with a small amount of one other element. The structure on the right represents the same metal alloyed with two other elements.

Fig 2.5

Chemists need pure substances. If a substance is pure it always has the same properties. It will always react in the same way. This is not so with an impure substance. Its properties can vary.

The properties and reactions of pure salt are always the same. The properties of impure salt depend on the type of impurity present and on its concentration.

Nearly all pure substances that chemists use have gone through two stages: they have been separated from a mixture and have been tested to find out how pure they are.

2.1 Separating mixtures

If we want to separate two substances in a mixture, we must find some difference between them. We use this difference to choose our method of separation.

1. Using differences in solubility

In a mixture of sand and water, the water particles are very small and will pass through a filter paper.

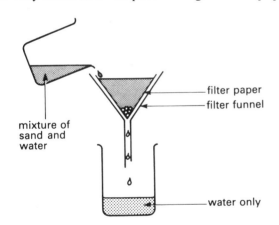

filter paper
filter funnel

mixture of sand and water

water only

Fig 2.6 Apparatus for filtration

The sand particles are much larger and will not pass through. We can therefore separate sand and water by filtration.

The *liquid* that passes through the filter paper is known as the *filtrate*.
The *solid* that stays on the filter paper is known as the *residue*.

The filter paper is a very fine sieve. Its holes are smaller than sand particles but bigger than water

particles. Filtration or sieving is often used as a method of separation.

Filtration cannot be used to separate two solids such as sand and salt, because the pieces of both substances are too big to go through the filter paper.

We can make the salt particles much smaller by adding water. The salt dissolves in water forming very tiny particles which can pass through the very small holes in the filter paper. If this mixture is now filtered, the sand will remain on the filter paper and the filtrate will be a solution of salt.

2. Using differences in boiling point

Suppose you wanted to get pure water or pure salt from salt water. Filtration would be no use, since dissolved salt particles are about the same size as water particles. Another difference in property must be used. Sodium chloride melts at 806 °C and water boils at 100 °C. This difference can be used and the mixture separated.

Evaporation

If you just want to get the salt from the solution, you can boil the salt water in an evaporating basin.

heat

Fig 2.7

The water has a lower boiling point. It will boil away leaving solid salt. (A bunsen flame is not hot enough to boil sodium chloride.)

Only evaporate all the water away if the solid is unaffected by heat. Sugar for example, will break down when heated strongly. If you want to obtain sugar from a sugar solution, heat gently until most of the water has boiled away. Then leave the remaining solution to cool. Crystals will be obtained and these can be filtered off from the remaining liquid.

Distillation

If you need to get pure water from salt water, then you need to use distillation (see Fig 2.8).

In the distillation flask the water is changed into water vapour, this passes into the condenser, where it is cooled and changes back into water. The condenser is sloped so that the pure water formed runs away from the distillation flask into the collecting beaker.

If we wish to separate two liquids with different boiling points (eg: ethanol, boiling point 80 °C, and water, boiling point 100°C) by distillation, *fractional distillation* must be used.

The fractionating column is a long tube packed with glass beads, these are to make the process work better.

The thermometer bulb must be level with the condenser so that the boiling point of the liquid being collected is measured. The liquid with the lower boiling point is collected first. When no more passes into the condenser, the beaker is changed and the flask is heated more strongly to collect a substance with a higher boiling point. In this way the original liquid mixture is divided up into a number of *fractions* in separate beakers.

One of the major uses of fractional distillation is in refining crude oil. You can see (page 118) how industry uses fractional distillation when refining crude oil.

This process is also used to obtain ethanol from fermented liquor (page 127) and nitrogen and oxygen from liquid air (page 88).

3. Separating substances that sublime

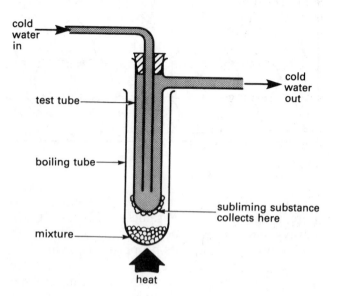

Fig 2.10 Separating by subliming

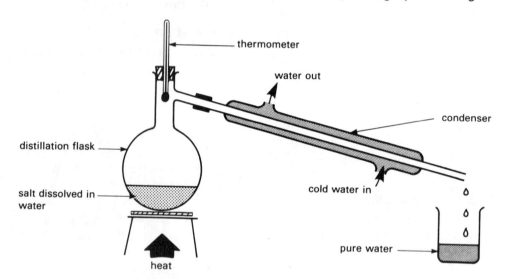

Fig 2.8 Laboratory apparatus for distillation of salt water

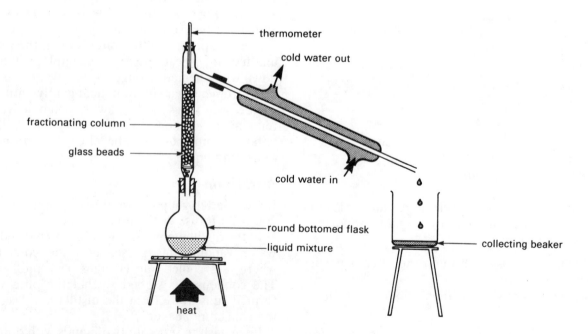

Fig 2.9 Laboratory apparatus for fractional distillation

Some substances, such as iodine, do not usually melt or boil when heated. These substances *sublime* (change directly from a solid to a gas). This property can be used to separate substances that sublime from those that do not (see Fig. 2.10).

When the mixture is heated, the substance that sublimes changes to a gas. This rises up the tube until it reaches the test tube that is water-cooled. On the surface of the water-cooled test tube the substance changes back from a gas into a solid.

4. Chromatography

Chromatography is a very useful technique. It can be used to separate tiny quantities of very similar substances.

A small spot of the mixture in solution is placed on the filter paper strip and the end of the strip is dipped into the solvent. As the solvent moves up the filter paper, it pulls the substances to be separated along with it. The more soluble a substance is in the solvent, the more it will keep up with the solvent as it soaks up the paper. This separates the mixture. Fig 2.12 shows what is seen when three dyes are separated.

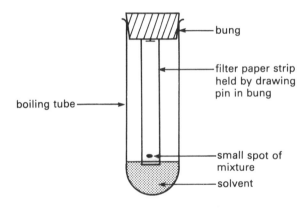

Fig 2.11 Apparatus for simple paper chromatography

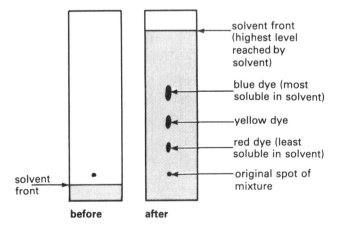

Fig 2.12

The separate spots of dye can then be cut out and the dyes washed into solution using suitable solvents.

The separation of substances by this method is clearly seen if they are coloured. Colourless substances can also be separated as long as we have some way of finding out where the spots are.

Chromatography can be used to find out what flavourings have been added to food, or even to prove that a sample of engine oil has come from a particular car.

Experiment 2.1 Separating indicators by chromatography

Many experiments can be carried out to separate colouring matter in inks using paper chromatography with a suitable solvent. (Felt tip pens are another source of coloured mixtures.)

Take a large rectangular piece of filter paper or chromatography paper and draw a pencil line 2 cm from one end. Using a drawn out melting point tube, put small drops of a number of indicators 3 cm apart along the pencil line. Indicators such as methyl red, methyl orange, screened methyl orange, phenolphthalein and Universal Indicator are useful.

The solvent (butanol, ethanol, concentrated aqueous ammonia and water in the ratio 24:8:1:7 by volume) should be poured into a suitable container.

The solvent level must initially be below the level of the spots. Stand the paper in the solvent and put a cover on the vessel. When the solvent has nearly reached the top of the paper, carefully take it out and allow it to dry. Mark any spots with a pencil (see Figs. 2.11 and 2.12).

Put the paper near (i) an open bottle of aqueous ammonia, (ii) an open bottle of concentrated hydrochloric acid to see if any further spots can be detected.

You should now be able to decide how many substances are in each of the indicators and also what indicators are used to make Universal indicator.

2.2 Knowing it is pure

Once a substance has been purified it must be tested to prove that it is pure. It is worth considering some of the properties of pure substances.

Properties of pure substances

1. It will look pure. *Example*: A white substance cannot be pure if it has black specks in it.
2. It will have a fixed melting point and boiling point. *Example*: Water melts at 0 °C and boils at 100 °C at atmospheric pressure.
3. It cannot be separated into anything simpler by methods such as filtration, distillation, chromatography, crystallisation or fractional distillation. *Example*: A pure red dye only gives one spot on a chromatogram. A mixture of red dyes gives several.

By investigating one or more of these properties, we can find out if a substance is pure.

A method often used is to measure the melting point of the substance.

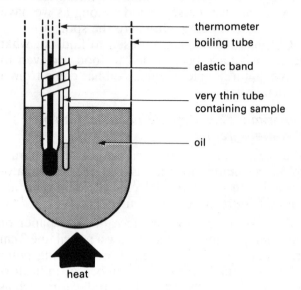

Fig 2.13

A small sample of the substance is placed in a very thin tube. This is strapped to a thermometer with the sample level with the bulb of the thermometer. The thermometer and sample are placed in an oil bath and the oil is *slowly* heated. The sample in the thin tube is watched carefully and the temperature at which it melts is noted.

If the substance is pure it melts completely at one temperature. If it is impure it melts over a temperature range.

Purity of substances is very important. We must be certain that food and medicines do not contain harmful substances. Very small amounts of an impurity could cause death. The food and drugs industries must check constantly to make sure that the substances they use are pure.

Questions

1. Which one of the following methods is the best way to obtain crystals of sodium nitrate from aqueous sodium nitrate?

A chromatography
B distillation
C evaporation
D filtration
E sublimation

2. Which one of the following mixtures can be separated by adding water, stirring and filtering?

A calcium carbonate and sodium chloride
B carbon and iron
C hydrochloric acid and nitric acid
D nitrogen and oxygen
E sodium chloride and sugar

3. Which one of the following properties shows that a liquid is pure?

A It does not mix with water.
B It is colourless and odourless.
C It has no affect on litmus paper.
D It boils at a fixed temperature at a given pressure.
E It passes unchanged through a filter paper.

4. A chemist suspects that a mixture contains traces of a green dye (boiling point 70 °C) as well as orange and red dyes (boiling points 69 °C and 70 °C respectively). Which one of the following is the best method by which the green dye may be separated?

A evaporation of the water present
B filtration
C fractional distillation
D paper chromatography
E using a separating funnel

5. In order that substances may be separated by paper chromatography, it is necessary that:

A they are both liquids
B they are both soluble in the same solvent
C they have different densities
D they have different colours
E they have different boiling points.

6. Describe fully what each of the following diagrams represents:

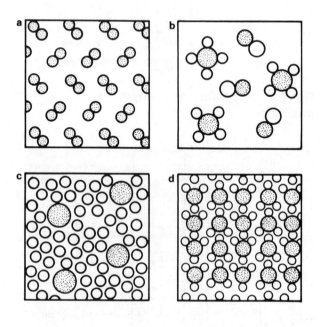

7. In which of the following is a change of state taking place?

a toasting bread
b melting ice
c boiling potatoes
d buttering bread
e dissolving sugar in water
f burning coal
g boiling water
h frying an egg

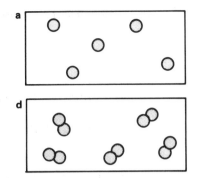 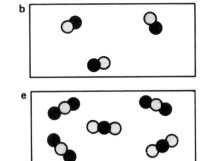 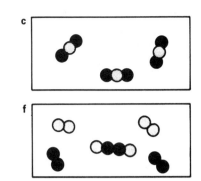

8.

    ○ represents atoms of element A and

    ● represents atoms of element B.

Decide whether each of the above boxes contains an element, a compound or a mixture

9. CHROMATOGRAPHY, CRYSTALLISATION, DISTILLATION, FRACTIONAL DISTILLATION, FILTRATION.

Choose from the above techniques the one most suitable for the following separations.

a Getting pure water from sea water
b Getting salt crystals from salt water
c Removing the dust from air
d Separating black ink into its individual dyes
e Getting turpentine from turpentine contaminated with paint
f Getting pure water from a cup of tea

10.

a With the aid of a diagram, describe how you would determine the melting point of a solid.
b State how your result would indicate whether the solid was a pure substance or a mixture of different substances.

[C]

11. The diagram below shows the chromatogram produced by three dyes and four different inks.

a Which dye is not a pure substance?
b What colour would you expect ink 1 to be?
c Which ink is green?
d Which two inks are identical?
e Which dye would you expect to be least soluble in the solvent used? Give a reason for your answer

12.

a You are provided with a powdered mixture of calcium carbonate, sodium chloride and naphthalene (a solid hydrocarbon).
  (i) Describe how, by the use of suitable *named* solvents, you could obtain pure samples of *each* of the three constituents of the mixture.
  (ii) Outline a method by which you could determine the percentage by mass of naphthalene in the mixture.

b A given sample of a compound is known to have a melting point between 60 °C and 80 °C but the exact melting point of the pure compound is unknown. Explain briefly how you could find out whether the sample is pure or impure.

[C]

13.

a You are provided with a solution of an organic compound A in a liquid hydrocarbon B. The melting point of A is approximately 120 °C and the boiling point of B is approximately 70 °C.
  (i) Describe, including a diagram of your apparatus, how you would prepare a pure sample of B and a good crystalline sample of A.
  (ii) If the exact melting point of A is not known, how would you test the purity of your sample of A?

b Describe an experiment you have seen or performed to illustrate the separation of substances by chromatography

[C]

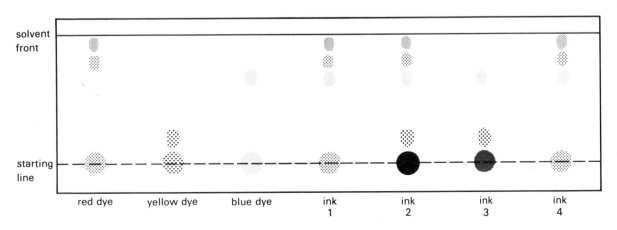

3 Atoms — the simplest particles

Elements are the simplest substances, from which all other substances are made. The smallest particle of an element that can exist is known as an *atom*.

Atoms are very, very small. They have a diameter of about one ten millionth of a centimetre. Look at a one centimetre length on your ruler. Try to imagine ten million particles fitting into that space.

Atoms are far too small to be seen, so scientists have had to imagine a model of the atom. As scientists get more information about atoms, they gradually improve their model.

At first it was thought that atoms were hard balls. Later it was found that atoms could be broken down by strong electric fields into charged particles. Later still, experiments suggested that atoms were not solid, but contained large empty spaces. These and other discoveries led to the model of the atom that we accept today.

3.1 Structure of the atom

Atoms are now thought to be made up of three different particles; protons, neutrons and electrons. Each atom is made up of two regions; the nucleus and the shells of electrons (Fig 3.1).

The nucleus: The nucleus is extremely small. It is about one ten thousandth of the diameter of the atom. To get some idea of this size, imagine an atom in which the nucleus was the same size as you. This atom would have a diameter of about ten miles.

The nucleus contains *protons* that are positively charged and *neutrons* that are uncharged. Nearly all the mass of an atom is due to the protons and neutrons. Therefore the nucleus is very dense.

The shells (energy levels): The shells contain negatively charged *electrons*. These spin round the nucleus like planets round the Sun. Electrons are very light, even compared with protons and neutrons. 1840 electrons weigh the same as 1 proton.

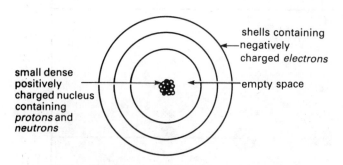

Fig 3.1 The atom

Name of particle	relative mass	relative charge
Proton	1	+1
Neutron	1	0
Electron	almost 0	−1

Table 1

Table 1 shows some of the properties of protons, neutrons and electrons.

You can see from Table 1 that protons and electrons have the same charge, but this charge is opposite in sign. Since atoms are electrically neutral, they must *contain equal numbers of protons and electrons*. You should also see that the mass of an atom depends mainly on the number of protons and neutrons it contains.

We need to define two terms to help us describe the structure of atoms of different elements.

1. *Atomic number:* The atomic number of an element is the number of protons in the nucleus of an atom of that element. Each element has a different atomic number.

2. *Mass number:* The mass number is the sum of the number of protons and the number of neutrons in the nucleus of an atom.

A carbon atom that contains 6 protons and 6 neutrons has a mass number of 12. A sodium atom that contains 11 protons and 12 neutrons has a mass number of 23.

3.2 Isotopes

For many years it was thought that all atoms of the same element were identical. It was eventually realised that atoms of the same element could have different masses. This is because they contain different numbers of neutrons in their nuclei.

Isotopes are atoms of the same element having different mass numbers. They therefore have the same number of protons but a different number of neutrons. All isotopes of the same element have the same chemical properties.

Chlorine is an element which has isotopes. All chlorine atoms have an atomic number of 17. This means that all chlorine atoms contain 17 protons and 17 electrons. The two main isotopes of chlorine have mass numbers of 35 and 37. Therefore:

The isotope of mass number 35 must contain $35 - 17 = 18$ neutrons.

The isotope of mass number 37 must contain $37 - 17 = 20$ neutrons.

Chemists use a shorthand form to describe the isotopes of elements. They write the mass number and the atomic number before the symbol for the element.

$$\text{mass number} \atop \text{atomic number} \quad \text{SYMBOL}$$

The isotopes of chlorine are therefore written as $^{35}_{17}Cl$ and $^{37}_{17}Cl$.

3.3 Relative atomic mass (A_r)

The mass number of an isotope tells us approximately how heavy an atom of that isotope is. However, there is a more accurate scale which is used to measure the masses of atoms. This is the relative atomic mass scale. The isotope of carbon of mass number 12 ($^{12}_6C$) is taken as the standard on this scale.

It is given a relative atomic mass of 12. The masses of all other atoms are compared with this. If an atom were twice as heavy as an atom of $^{12}_6C$, it would have a relative atomic mass of $2 \times 12 = 24$. If the atoms were five times as heavy it would be $5 \times 12 = 60$.

Most elements have isotopes and most samples of an element contain a mixture of isotopes. The relative atomic mass scale allows for this. The following example shows how.

In a normal sample of chlorine, three-quarters of the atoms are of the $^{35}_{17}Cl$ isotope and one-quarter are of the $^{37}_{17}Cl$ isotope.

The relative atomic mass of chlorine is therefore:
$(\frac{3}{4} \times 35) + (\frac{1}{4} \times 37) = 35 \cdot 5$.

A table showing the relative atomic masses of the elements is given at the end of the book.

3.4 What about the electrons?

You might say "Electrons are so small, they can't have much effect on the properties of an atom." Nothing could be further from the truth. Chemists believe that electrons control the chemical properties of substances. Different substances have different chemical properties because they have different numbers of electrons and because the electrons are arranged differently in the shells.

Fig 3.2 The hydrogen atom

Hydrogen is the least dense element. It has the simplest atoms. Hydrogen has an atomic number of 1 and therefore only one electron. This electron goes into the shell nearest the nucleus. This is known as the first shell (energy level).

The first shell can hold a maximum of two electrons, so in the lithium atom (atomic number 3) one electron has to go into the second shell.

Fig 3.3 A lithium atom

The second shell can hold a maximum of eight electrons. This means that sodium with an atomic number of eleven is the first element to have electrons in the third shell.

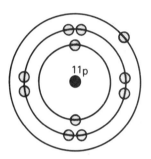

Fig 3.4 A sodium atom

Diagrams showing the arrangement of electrons in an atom use a lot of space. Chemists use a shorthand form. This just shows the number of electrons in each shell without drawing the shells.

Hydrogen has an electron arrangement 1
Lithium has an electron arrangement 2,1
Sodium has an electron arrangement 2,8,1

Table 2 shows the electron arrangement of elements with atomic numbers 1–20.

Name of element	Chemical symbol	Atomic number	Electron arrangement
Hydrogen	H	1	1
Helium	He	2	2
Lithium	Li	3	2,1
Beryllium	Be	4	2,2
Boron	B	5	2,3
Carbon	C	6	2,4
Nitrogen	N	7	2,5
Oxygen	O	8	2,6
Fluorine	F	9	2,7
Neon	Ne	10	2,8
Sodium	Na	11	2,8,1
Magnesium	Mg	12	2,8,2
Aluminium	Al	13	2,8,3
Silicon	Si	14	2,8,4
Phosphorus	P	15	2,8,5
Sulphur	S	16	2,8,6
Chlorine	Cl	17	2,8,7
Argon	Ar	18	2,8,8
Potassium	K	19	2,8,8,1
Calcium	Ca	20	2,8,8,2

Table 2

Some elements are very similar to others. Sodium, potassium and lithium have very similar chemical reactions. Notice that they each have one electron in their outermost shell.

Chlorine and fluorine have very similar chemical reactions. Notice that they both have seven electrons in their outermost shell. Chlorine and fluorine are very different from lithium, sodium or potassium. It seems likely that the electrons in the outermost shell are the ones mainly responsible for chemical properties.

The outermost shell electrons are known as *valency* electrons. Sodium, lithium and potassium each have one valency electron. Chlorine and fluorine each have seven valency electrons.

Scientists now have a very detailed model of the atom. This has slowly developed over the past hundred years. It will continue to develop. One day we might find out more about the structure of the nucleus. Perhaps protons, neutrons and electrons themselves can be broken down. Scientists will continue to make their model as accurate as possible, using all the information they can find.

Questions

1. Which one of the following statements is true for all atoms?

A number of electrons = number of neutrons
B number of electrons = number of protons
C number of neutrons = number of protons
D number of neutrons = number of protons + number of electrons
E number of electrons = number of protons + number of neutrons

2. A neutral atom of magnesium (atomic number 12, mass number 24) contains:

A 12 protons, 12 neutrons and 12 electrons
B 24 protons, 12 neutrons and 12 electrons
C 12 protons, 12 neutrons and 24 electrons
D 12 protons, 12 neutrons and 24 electrons
E 12 protons, 24 neutrons and 24 electrons.

3. From the symbol $^{7}_{3}Li$ it can be deduced that:

A an atom of lithium has three protons and seven neutrons in its nucleus
B an atom of lithium contains three electrons
C lithium has a mass number of 3
D lithium has an atomic number of 7
E the electronic structure of lithium is 2, 5.

4. Which one of the following is the arrangement of electrons in an atom of $^{16}_{8}O$?

A 2, 4, 2
B 2, 6
C 4, 4
D 2, 8, 6
E 2, 8, 4, 2

5. A phosphorus atom contains 15 protons: which one of the following statements is true?

A The atom contains 7 electrons in the outer shell.
B The atom contains 15 neutrons.
C Phosphorus is a metal.
D A_r of phosphorus is 30
E The atomic number of phosphorus is 15.

6. ELECTRONS, ION, ISOTOPE, PROTONS, NEUTRONS, NUCLEUS. Choose words from the above list to complete the following passage. Each word can be used once, more than once, or not at all.

All atoms are made up of small, dense (1) ____ surrounded by shells containing (2) ____. Most atoms contain three different particles: protons, (3) ____ and electrons. (4) ____ are positively charged, (5) ____ are negatively charged and (6) ____ are uncharged. The lightest of these particles are the (7) ____. All atoms of the same element contain the same number of (8) ____ and (9) ____. Atoms of the same element with different numbers of (10) ____ are known as isotopes.

7. Draw diagrams to show the structures of the atoms represented by:
a $^{12}_{6}C$ b $^{23}_{11}Na$ c $^{35}_{17}Cl$ d $^{1}_{1}H$

8. Use the information shown in the Table to answer the following questions.

Name of particle	Number of protons	Number of neutrons	Number of electrons
A	17	18	17
B	11	12	10
C	10	10	10
D	8	8	10
E	17	20	17
F	12	12	12
G	16	16	16

a Which particle has more protons than electrons?
b Which particle has more electrons than protons?
c Which is the heaviest particle?
d Which is the lightest particle?
e Name two particles that are neutral atoms
f Which of the particles is an atom with two electrons in its outermost shell?
g Which of the particles is a neutral atom with six electrons in its outermost shell?
h Which of the particles is negatively charged?
i Which of the particles is positively charged?
j Which of the particles are isotopes of the same element?

4 The way in which atoms combine

When two atoms collide, they can either stick together, or bounce apart. They either combine or they do not.

The only substances that exist as individual atoms are the noble gases (helium, neon, argon, xenon, krypton and radon). All other substances exist as clusters of atoms stuck together.

In the element hydrogen, pairs of atoms stick together.

In the element sulphur, groups of 8 atoms stick together.

In the element carbon, millions of atoms stick together.

When *atoms* of *different* elements stick together a *compound* is formed.

4.1 Why do atoms combine?

The chemical properties of an element depend on the arrangement of electrons in the atoms of that element. The noble gases are monatomic (exist as single atoms) and are unreactive because they have a very stable arrangement of electrons. If two atoms of a noble gas collide, they will not combine. They will bounce apart. However, the atoms of all other elements have less stable electron arrangements. They can combine when they collide. When atoms combine they get the stable electron arrangement of a noble gas and hence become more stable.

4.2 How do atoms combine?

Atoms can obtain a more stable electron arrangement in a number of different ways.

1. *Electron transfer*

Sodium has an atomic number of 11. A sodium atom therefore has 11 protons and 11 electrons. The protons are in the nucleus and the electrons are in shells around the nucleus as shown in Fig 4.1.

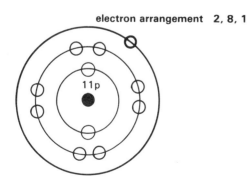

electron arrangement 2, 8, 1

Fig 4.1 The sodium atom

The sodium atom gets the stable arrangement of a neon atom (2,8) by losing its outermost electron. When this happens it is left with 11 protons (11+ charges), but only 10 electrons (10− charges). It is no longer a neutral atom. It is positively charged. The sodium *atom* changes into a sodium *ion* (1 + charge).

An ion is a charged atom or cluster of atoms.

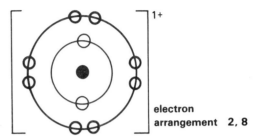

electron arrangement 2, 8

Fig 4.2 The sodium ion

Chlorine has an atomic number of 17. The chlorine atom therefore has 17 protons and 17 electrons as shown in Fig 4.3.

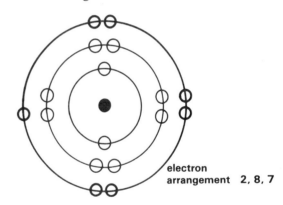

electron arrangement 2, 8, 7

Fig 4.3 The chlorine atom

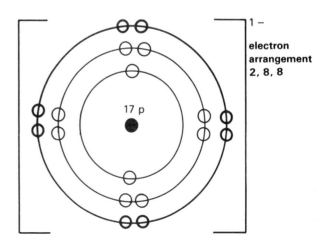

electron arrangement 2, 8, 8

Fig 4.4 The chloride ion

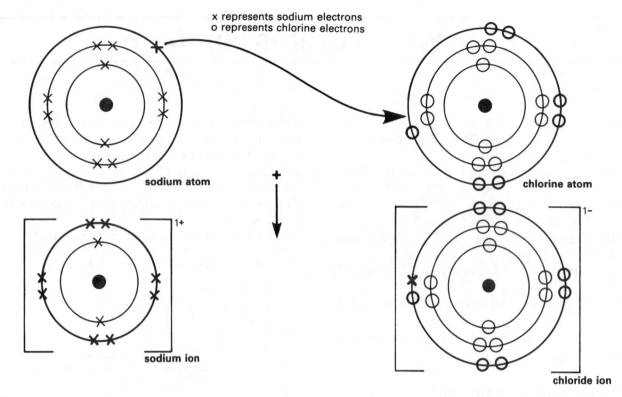

x represents sodium electrons
o represents chlorine electrons

Fig 4.5 Forming sodium chloride

The chlorine atom gets a stable electron arrangement by finding one more electron for its outermost shell. When this happens it gets the stable electron arrangement of an argon atom (2,8,8). The chlorine atom is left with 17 protons (17+ charges) and 18 electrons (18– charges). It is a negatively charged ion (1– charge). The ion formed when a chlorine atom gains an electron is known as the *chloride* ion.

The sodium atom needs to lose an electron to become stable. The chlorine atom needs to gain an electron to become stable. If a sodium atom collides with a chlorine atom it is possible for both atoms to become stable at the same time. This is shown in Fig 4.5.

By sodium giving its outermost electron to the chlorine atom, both atoms get a stable electron arrangement. The compound sodium chloride is formed. Sodium chloride is made up of ions. It is known as an *ionic compound*.

Since sodium ions and chloride ions are oppositely charged, they attract each other. As the ions stick together a crystal of sodium chloride is built up as shown in Fig 4.6

The attractive forces that hold the ions together in an ionic solid are known as *ionic bonds* (electrovalent bonds).

Magnesium oxide is another substance that is formed by *transferring electrons*.

Magnesium (atomic number 12) has 12 protons and 12 electrons. It has the electron arrangement 2,8,2. To form a stable electron arrangement it must lose 2 electrons.

Oxygen (atomic number 8) has 8 protons and 8 electrons. It has an electron arrangement 2,6. It needs to gain 2 electrons to get a stable electron arrangement.

By electron transfer, magnesium atoms form magnesium ions and oxygen atoms form oxide ions. This is shown in Fig 4.7, see page 19.

You should note that the ions in magnesium oxide have double the charge of the ions in sodium chloride.

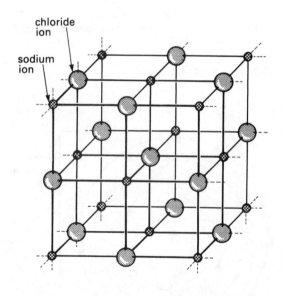

Fig 4.6 The structure of sodium chloride

18

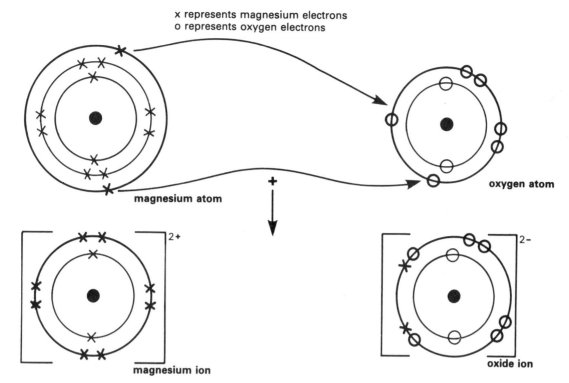

Fig 4.7 Forming magnesium oxide

2. *Electron sharing*

Some atoms get a stable electron arrangement by sharing. Consider hydrogen: a single hydrogen atom has just one electron. This is unstable. If two hydrogen atoms join together, it is possible for electrons to be shared. This is shown in Fig 4.8.

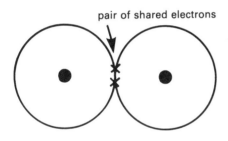

Fig 4.8

By sharing electrons, each hydrogen atom has *two* electrons in its shell. Each atom therefore has the stable electron arrangement of a helium atom.

The shared pair of electrons is known as a *covalent bond*. Covalent bonds hold atoms together very strongly.

By sharing electrons the two hydrogen atoms form a hydrogen *molecule*. Most molecules consist of a group of atoms held together by covalent bonds. Molecules are neutral. They are not charged like ions. A substance that is made up of molecules is known as a covalent substance.

Many non-metal elements exist as molecules. Chlorine is another example (see Fig 4.9 below).

Many compounds are also formed by sharing electrons. Methane (CH_4) is an example. In methane four pairs of electrons are shared. Four covalent bonds are formed. By forming these bonds each hydrogen atom gets two electrons in its electron shell. Each gets the stable electron arrangement of a helium atom. Carbon atoms get eight electrons in their outer shells. They get the

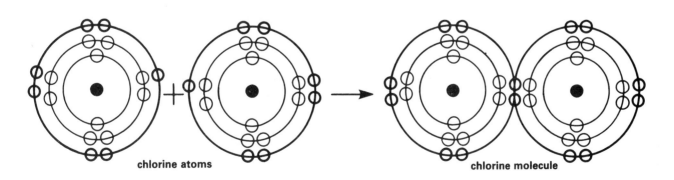

Fig 4.9 Forming chlorine molecules

stable electron arrangement of neon atoms (2,8). This is shown in Fig 4.10.

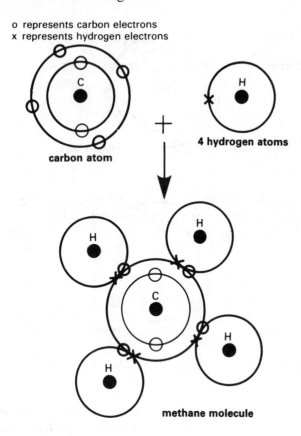

o represents carbon electrons
x represents hydrogen electrons

carbon atom

4 hydrogen atoms

methane molecule

Fig 4.10 Forming methane

When a covalent compound is formed, only the valency electrons are involved in forming covalent bonds. Because of this we can simplify the picture of a methane molecule to that shown in Fig 4.11.

$$
\begin{array}{c}
\text{H} \\
\overset{\text{x o}}{} \\
\text{H} \ \overset{\text{o}}{\underset{\text{x}}{}} \ \text{C} \ \overset{\text{x}}{\underset{\text{o}}{}} \ \text{H} \\
\overset{\text{o x}}{} \\
\text{H}
\end{array}
$$

Fig 4.11

In Fig 4.11 the shells are left out and only the electrons that form covalent bonds are shown. The picture of the methane molecule can be shown more simply by showing each covalent bond as a line, instead of as a pair of electrons.

The picture of the methane molecule then becomes:

$$
\begin{array}{c}
\text{H} \\
| \\
\text{H} - \text{C} - \text{H} \\
| \\
\text{H}
\end{array}
$$

Fig 4.12

A line is often used to represent a covalent bond when chemists want to show the structure of a molecule as simply as possible.

Example:

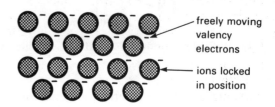

is water

is ethane

$$O = C = O$$ is carbon dioxide

3. Metallic bonding

Metals are usually dense, high melting point solids. This means that the metal ions must be packed closely together and there must be strong attractive forces between the ions.

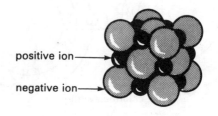

freely moving valency electrons

ions locked in position

Fig 4.13 The structure of metals

The structure can be thought of as positive ions in a sea of electrons. The positive ions repel one another, but they are held together by their attraction for the 'free' electrons. These electrons make metals good conductors of electricity.

4.3 Types of structure

Atoms can stick together by forming ions, by sharing electrons, or by giving electrons to a central pool, as in metals. In each case, only the valency electrons are involved. Because of the different types of bonding that can exist, matter can occur in one of four different structures:

1. Ionic giant structures

These are structures formed by ionic substances such as sodium chloride. A large number of ions pack together in a regular pattern to form a crystal structure.

positive ion

negative ion

Fig 4.14 Ionic giant structure

2. Metal giant structures

These structures are found in metals. Large numbers of ions pack together closely in a regular pattern to form a crystal structure.

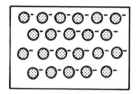

Fig 4.15 Metal giant structure

3. Covalent giant structures

These are found in elements (atoms are all alike) such as diamond and graphite. They are also found in compounds (more than one type of atom) such as silica (which is made up of silicon atoms and oxygen atoms). Large numbers of atoms stick together by forming covalent bonds to give a very strong crystal structure. The similarity in structure between diamond and silica (sand) gives these two substances similar properties. They are both very hard with high melting points and high boiling points.

Fig 4.16 Covalent giant structure

4. Molecular structures

These structures are found in elements and compounds. Noble gases, such as helium and neon, have one atom per molecule. Chlorine gas has two atoms of chlorine per molecule. Sulphur has eight atoms of sulphur per molecule.

These structures are also found in compounds. In water molecules there are three atoms per molecule. In butane molecules there are fourteen atoms per molecule. The atoms within any molecule are very tightly held, but the attractive forces between different molecules are usually weak. Most covalent compounds have molecular structures.

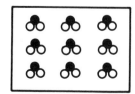

Fig 4.17 Molecular structure

4.4 The effect of bonding on properties

Compounds that are formed in different ways have different properties. Ionic compounds have different properties from covalent compounds.

Experiment 4.1 Bonding in compounds

You will need: lead(II) bromide, naphthalene, paraffin wax, potassium carbonate, sodium chloride and sugar. Carry out the following tests on each of the compounds.

1. Heat a little of the solid in a test tube. Stop heating as soon as the solid has melted or after about one minute.
2. Test the solubility of each solid in water. If the solid dissolves, keep this solution for test **3**.
3. Using the apparatus shown in Fig 10.4 (page 65), test each solution from **2.** to see if the aqueous solution conducts an electric current.

Can you divide these six compounds into ionic compounds and covalent compounds? The differences between ionic compounds and covalent compounds is summarised in Table 1 overleaf.

4.5 Bonding and the Periodic Table

The Periodic Table (Chapter 6) can act as a useful guide to the type of compounds that elements form.

Metals form ionic compounds in which the metal atom has been changed to a metal ion. Metal ions are positively charged. The number of charges on the metal ion is usually equal to the group number of that element.

Sodium is in Group I. It forms the ion Na^+.
Aluminium is in Group III. It forms the ion Al^{3+}.

Non-metal elements form ionic compounds and covalent compounds. When a non-metal element reacts with a metal it usually forms an ionic compound. In these compounds the non-metal atom is changed to a negative ion. When a non-metal element reacts with another non-metal element it forms a covalent compound.

Questions

1. Which of the following represents an ion with a charge of 2+?

	protons	neutrons	electrons
A	2	4	2
B	8	8	8
C	4	5	4
D	12	12	10
E	16	16	18

2. The atomic number of sodium is 11. When sodium reacts to form an ionic compound, the electronic configuration of the sodium ion formed is:

A 2, 7   D 2, 8, 2
B 2, 8   E 2, 8, 3.
C 2, 8, 1

Properties of ionic compounds eg sodium chloride

1. They are crystalline solids	Positive and negative ions attract each other. They build up into a regular crystal structure.
2. Have high melting points and boiling points	When an ionic substance melts each ion is free to move. A lot of energy (heat) is needed to overcome the attractive forces between ions. Metal oxides, such as magnesium oxide, have very high melting points. They are used as refractory materials, i.e. for lining furnaces.
3. Are usually soluble in water but insoluble in solvents such as tetrachloromethane	Water molecules attract the ions. The water molecules separate the ions and keep them apart.
4. Conduct electricity when molten or dissolved in water	If charged particles can move, electricity can flow. When ionic substances are molten or dissolved the ions are free to move.

Properties of covalent compounds eg naphthalene

1. Usually have low melting points and boiling points—so they are often gases or liquids or soft solids	The atoms within a molecule are strongly held together, but there are only weak attractive forces between molecules. The molecules are easily separated.
2. They do not conduct electricity	Molecules are neutral. Charged particles are needed to conduct electricity.
3. Do not usually dissolve in water but are soluble in tetrachloromethane	Water molecules cannot attract the molecules of the covalent substance strongly enough to separate them. Covalent giant structures are solids and are difficult to dissolve in most solvents.

Table 1. Comparing the properties of ionic compounds with covalent compounds

3.

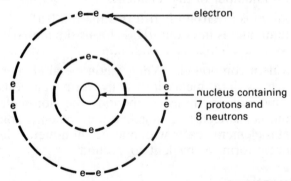

Given that the atomic numbers of the elements: nitrogen, oxygen, neon, chlorine and argon are 7, 8, 10, 17 and 18 respectively, the symbol for the particle shown above is:

A Ar   D Ne
B Cl^- E O^{2-}
C N^{3-}

4. The diagram opposite illustrates the structures of the atoms of two elements X and Y.

When these two elements combine together they form the compound:

A X_2Y D X_2Y_3
B XY E X_3Y_2
C XY_2

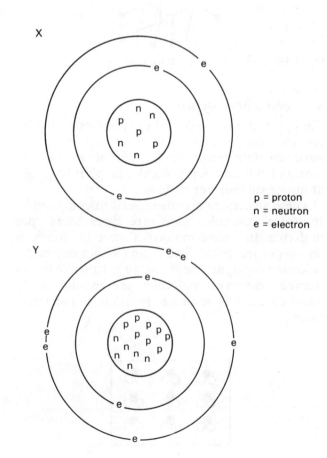

p = proton
n = neutron
e = electron

5. An element M has the electronic structure 2, 8, 5. The simplest compound formed when M combines with hydrogen is most likely to be:

A a solid non-electrolyte
B a solid that conducts an electric current when it is melted
C a liquid that conducts electricity
D a gas that dissolves in water to form a solution that conducts an electric current
E a gas that dissolves in organic solvents to form a conducting solution.

6. Sodium atoms contain 11 electrons, chlorine atoms contain 17 electrons.

a Draw diagrams to show the arrangement of electrons in:
(i) a sodium atom, (ii) a chlorine atom.
b Draw a diagram to show the arrangement of electrons, and the type of bonding in sodium chloride.

7. Nitrogen atoms contain 7 electrons

a Draw a diagram to show the arrangement of electrons in a nitrogen atom.
b Draw a diagram to show the arrangement of electrons and the type of bonding in ammonia (NH_3).

8. Draw diagrams to show the arrangement of electrons in the following particles. Look up the atomic number of each element on the data page.

a a magnesium atom
b a sodium ion (Na^+)
c an oxide ion (O^{2-})
d a carbon atom
e a hydrogen molecule

9. ARGON, ALUMINIUM, DIAMOND, MAGNESIUM OXIDE, METHANE, SODIUM CHLORIDE, WATER.
From the above list choose:

a 2 ionic substances
b 2 substances that exist as molecules
c a substance that exists as a covalent giant structure
d a substance that exists as individual atoms
e a substance that conducts electricity when solid
f the substance with the highest melting point
g the substance with the lowest boiling point.

10. Substance A has a melting point of 804 °C. It dissolves in water to form a colourless solution.
This solution conducts electricity.
Substance B has a melting point of 1728 °C. It is insoluble in water. It does not conduct electricity.
Substance C has a melting point of 1083 °C. It is a good conductor of electricity at room temperature.
Substance D melts at −117 °C and boils at 78 °C. It dissolves in water to form a colourless solution. This solution does not conduct electricity.
Use this information to answer the following questions:

a which substance could be a metal element?
b only one of the substances smells—which one?
c which one of the substances is an ionic solid?
d which one of the substances is made up of molecules?
e which one of the substances is a covalent giant structure?

11. The diagram shows the arrangement of electrons in a methane molecule.

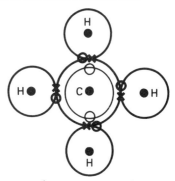

a how many electrons are there in a carbon atom?
b how many atoms are there in a methane molecule?
c how many covalent bonds are there in a methane molecule?
d how many electrons are 'shared' in a methane molecule?
e is a methane molecule, positively charged, negatively charged or neutral?

12. Make use of the information in the table below to answer the questions which follow.

Element	Lithium	Beryllium	Carbon	Oxygen	Aluminium	Chlorine
Symbol	Li	Be	C	O	Al	Cl
Atomic number	3	4	6	8	13	17
Relative atomic mass	7	9	12	16	27	35·5

a The electronic structure of the carbon atom may be written C 2, 4. Show in a similar way the electronic structures of the other five elements in the table.

b Which *three* elements in the table are conductors of electricity?

c Predict the formula of the simplest compounds formed:
  (i) between lithium and oxygen
  (ii) between carbon and chlorine
  (iii) between beryllium and oxygen.

d Which, if any, of the compounds in your answer to **C** would you expect to be liquid at room temperature and pressure? Give reasons for your answers.

e Give the numbers of neutrons present in one atom:
  (i) of beryllium
  (ii) of carbon
  (iii) of aluminium.

f The relative atomic mass of chlorine is not an integer (whole number). How can this fact be explained?

[C]

13.

a What do you understand by the terms:
  (i) atomic number
  (ii) mass number
  (iii) isotopes?

b Using the two isotopes of chlorine, $^{35}_{17}Cl$ and $^{37}_{17}Cl$, as your example, explain why:
  (i) the two isotopes have identical chemical properties, and
  (ii) the relative atomic mass of chlorine is not a whole number.

c How do the types of bonding between:
  (i) sodium and chlorine, and
  (ii) hydrogen and chlorine

show the significance of the noble gas electron arrangement in bonding theory?

[JMB]

14.

a Give the charges and approximate relative masses of electrons, protons and neutrons. Where in the atom are these particles found?

b Magnesium oxide is an ionic compound whereas carbon dioxide is a covalent compound.
  (i) By means of suitable electron diagrams, illustrate the bonding in these two compounds.
  (ii) Give *two* general properties which are associated with compounds containing ionic bonds. How do these properties differ in compounds containing covalent bonds?

c The atomic number of the noble gas krypton is 36. For the element of atomic number 34, predict:
  (i) whether the element is metallic or non-metallic,
  (ii) the charge on the ion that the element forms.

  Briefly explain your answers.

[C]

15. State *four* general types of structure taken up by solid elements and compounds. Give *two* examples of each.

Use the examples you have given to discuss the important physical properties possessed by substances with each of these structures. Draw a diagram to show how the particles are arranged in each of the four structures.

[L]

16. The numbers of protons, neutrons and orbital electrons in particles A to F are given in the following table:

Particle	Protons	Neutrons	Electrons
A	3	4	2
B	9	10	10
C	12	12	12
D	17	18	17
E	17	20	17
F	18	22	18

a Choose from the table the letters that represent:
  (i) a neutral atom of a metal
  (ii) a neutral atom of a non-metal
  (iii) an atom of a noble gas (inert gas)
  (iv) a pair of isotopes
  (v) a cation (positive ion)
  (vi) an anion (negative ion).

b Give the formulae of the compounds you would expect to be formed between: (i) C and D (ii) E and hydrogen (iii) C and oxygen. (You may use the letters C, D and E as symbols for the elements.)

  In *each* case, say whether the compound will be an acid, a base or a salt.

c What will be the formula of a compound containing only particles A and B? Give *two* physical properties you would expect this compound to have.

[C]

17.

a Write out and complete the following table:

	Mass number	Atomic number of element	Numbers of electrons	protons	neutrons
an aluminium **atom**	27	13			
an oxide **ion**		8			8
a calcium **ion**				20	20
a lithium **atom**	7		3		

b Explain in terms of electrons: (i) the formation of calcium oxide (ii) why calcium and aluminium do not react together.

c In a sample of lithium, it was found that some of the lithium atoms had a mass number of 7 and the rest had a mass number of 6.
State: (i) the name given to atoms of the same element but of different mass number (ii) the numbers of each type of particle present in a lithium atom of mass number 6.

d The relative atomic mass of lithium is given as 6·9. What can you deduce from this fact?

[C]

5 The language of chemistry

A language is a way of passing information from one person to another. Chemistry has its own language. It is used by chemists to describe chemicals and chemical change as neatly as possible. Many languages such as English, French and German have a common alphabet. The same letters are used to make different words. Other languages such as Russian and Arabic have their own alphabet. Like Russian and Arabic, the language of chemistry has its own unique letters which make up the chemical alphabet.

5.1 The chemical alphabet

The simplest chemicals that exist are the elements. Each element is given a shorthand form, known as its *chemical symbol*. These symbols are the letters of the chemistry alphabet. Table 1 shows the symbols of some common elements.

Name of element	Symbol
aluminium	Al
argon	Ar
barium	Ba
bromine	Br
calcium	Ca
carbon	C
chlorine	Cl
copper	Cu
chromium	Cr
iron	Fe
hydrogen	H
iodine	I
lead	Pb
lithium	Li
magnesium	Mg
manganese	Mn
nickel	Ni
nitrogen	N
oxygen	O
phosphorus	P
potassium	K
silicon	Si
silver	Ag
sodium	Na
sulphur	S
zinc	Zn

Table 1

In Table 1, many elements have a symbol which is the first, or first two letters of its name. *Example:* sulphur (S); calcium (Ca). Some elements have symbols which are totally different from the name. *Example:* sodium (Na); iron (Fe). The reason for this is that when these elements were given their symbols in the late 18th or 19th century they were known by their Latin names.

Before we can use the language of chemistry we must learn at least some of the alphabet.

The Periodic Table (see Chapter 6) gives the chemical symbols for each element. Use the Periodic Table to see how many words you can think of, which can be spelt using chemical symbols, in 10 minutes, What is the longest word that can be spelt using chemical symbols?

5.2 Formulae from symbols

The chemical symbols give chemists a shorthand way of describing atoms of different elements. To a chemist, 4H means 4 hydrogen atoms and 6Al means 6 aluminium atoms. We know that it is unusual for atoms to exist by themselves. They have a strong tendency to stick together. When atoms of different elements stick together compounds are formed. The language of chemistry has to be able to describe these compounds using chemical symbols. This description of the compound is known as the *formula* of the compound.

The compound water is made up of molecules. Each molecule of water contains 2 hydrogen atoms and one oxygen atom. We could write the formula of water as H2O1, where this means 2 hydrogen atoms and one oxygen atom. However this is confusing, (could it be 201 hydrogen atoms?). In order to make it less confusing we lower the numbers so that the formula can be written H_2O_1.

It seems pointless writing 1 every time 1 atom of an element is present; so the formula can be simplified to H_2O. If there is no number after the symbol for an element in a formula, this means that there is only 1 atom of that element in each *formula unit*.

Do you remember that there are two types of compounds: covalent and ionic? In covalent compounds the *formula unit* is the molecule. Therefore, the formula describes how many atoms of each type there are in one molecule of the compound. In ionic compounds no molecules exist. The compound is made up of ions. Therefore in ionic compounds the *formula unit* is the simplest cluster of ions that is electrically neutral.

Sodium chloride is made up of ions Na^+ and Cl^-. The formula of sodium chloride is NaCl.

Magnesium chloride is made up of ions Mg^{2+} and Cl^-. The formula of magnesium chloride must be $MgCl_2$.

POSITIVE			NEGATIVE		
1+	2+	3+	1−	2−	3−
Li Lithium	Mg Magnesium	Fe Iron(III)	OH Hydroxide	O Oxide	
Na Sodium	Ca Calcium	Al Aluminium	Cl Chloride	S Sulphide	
K Potassium	Ba Barium		Br Bromide	SO_3 Sulphite	
Ag Silver	Fe Iron(II)		I Iodide	SO_4 Sulphate	
H Hydrogen			NO_3 Nitrate	CO_3 Carbonate	
NH_4 Ammonium	Cu Copper(II)		NO_2 Nitrite		
	Zn Zinc		HCO_3 Hydrogen carbonate		
	Pb Lead(II)				

Table 2

The formula of a substance gives information about its 'make up' in a very convenient way.

Chlorine is a covalent gas. It is made up of molecules. In each molecule 2 chlorine atoms are stuck together. The formula of a chlorine molecule is Cl_2. If these two chlorine atoms were not stuck together they would be described differently ie 2Cl.

TNT is an example of a more complicated compound. TNT is a covalent compound with formula $C_7H_5N_3O_6$. Each TNT molecule must contain 7 carbon atoms, 5 hydrogen atoms, 3 nitrogen atoms and 6 oxygen atoms.

5.3 Writing formulae

For many years the chemical formulae of all compounds were found by experiment. For complicated substances like TNT experimental methods are still needed. However, the formulae of many simple compounds can be written directly from the name of the compound, using the idea of *valency* or oxidation number.

Valency can be thought of as a *number* which is given to an element or a group of elements. It allows the formulae of compounds to be written directly from their names.

Certain rules apply in writing formulae. These rules are:

1. The sum of the positive valencies must equal the sum of the negative valencies.
2. If any group, which contains more than 1 element, occurs more than once in a formula, it must be bracketed.

Example: The formula of calcium nitrate must be written as $Ca(NO_3)_2$ and not as $CaNO_{32}$.

The following examples show how the idea of valency can be used to write the formulae of compounds.

a. *Magnesium oxide*
Magnesium　　　Oxide
Mg　　2+　　O　　2−
The positive and negative valencies balance.
Therefore the formula is MgO.

b. *Potassium oxide*
Potassium　　　Oxide
K　　1+　　O　　2−
　　1+
　　——　　　——
　　2+　　　2−

Two potassium atoms are needed for the positive and negative valencies to balance.
Therefore the formula is K_2O.

c. *Aluminium oxide*
Aluminium　　　Oxide
Al　　3+　　O　　2−
Al　　3+　　O　　2−
　　　　　　O　　2−
　——　　　——
　6+　　　6−

Two aluminium atoms and three oxygen atoms are needed to balance the valencies.
Therefore the formula is Al_2O_3.

d. *Calcium hydroxide*
Calcium　　　Hydroxide
Ca　　2+　　OH　　1−
　　　　　　OH　　1−
　——　　　——
　2+　　　2−

One calcium atom needs two hydroxide groups for the valencies to balance.
Therefore the formula is $Ca(OH)_2$.

The bracket is needed because the 2 applies to both the oxygen and hydrogen atoms in the hydroxide group.

Oxidation number

In redox reactions (page 29) it is useful to consider the oxidation number of elements, particularly in compounds.

In many cases, the valency and the oxidation number are the same.

There are a number of simple rules we can use with oxidation numbers.

1. The oxidation number of any element in its free state is defined as being 0 (zero).

2. (i) For a compound, the sum of the oxidation numbers of the elements is zero.

(ii) For an ion, the sum of the oxidation numbers is the same as the charge on the ion.

3. In its compounds, oxygen shows an oxidation state of -2 and the Group I and Group II elements, such as sodium and magnesium, show oxidation states of $+1$ and $+2$ respectively.

If we look at some examples, you will see how simple it is.

(a) Magnesium oxide

As magnesium is in Group II it has an oxidation number of $+2$ and oxygen has an oxidation number of -2,

MgO $+2$ and $-2 = 0$

(This is the same as the first example on valency.)

(b) Magnesium chloride

$MgCl_2$ $+2$ and $2(-1) = 0$

(c) In most of its compounds, hydrogen shows an oxidation number of $+1$.

What then is the oxidation number of sulphur in H_2S?

$2(+1) +$ oxidation number of sulphur $= 0$

therefore the oxidation number of $S = -2$

(d) What is the oxidation number of manganese in MnO_2?

oxidation number of $Mn + 2(-2) = 0$

Therefore the oxidation number of $Mn = +4$

The name for MnO_2 is written as manganese(IV) oxide (read as "manganese four oxide"). The (IV) tells us that the oxidation state of manganese is $+4$.

In a similar way, the name copper(II) oxide tells us that the copper is in oxidation state $+2$. $KMnO_4$ is called potassium manganate(VII), the name tells us that the manganese is in oxidation state $+7$.

There is no need to include the oxidation number for an element which always shows the same oxidation state in its compounds.

5.4 Chemical equations

Possibly the most important job of the language of chemistry is to describe chemical reactions or chemical changes. It does this with chemical equations.

Magnesium burns brightly in oxygen to form magnesium oxide. This change can be written as a *word equation*.

magnesium + oxygen → magnesium oxide

The → means *reacts to form*.

If we replace the names of the chemicals with their formulae we get:

$Mg + O_2 → MgO$

This is a chemical equation for the reaction between magnesium and oxygen.

Looking at this equation it looks as if one oxygen atom disappeared as the reaction took place. There are 2 oxygen atoms on the left hand side of the equation and only 1 on the right hand side. This is impossible as *atoms cannot be created or destroyed in any chemical reaction*.

The chemical equation needs to be altered in some way. It needs to be *balanced*. The correct equation for the reaction between magnesium and oxygen is

$2Mg + O_2 → 2MgO$

A balanced equation must contain the same number of atoms of each element on both sides of the arrow. Equations are balanced by placing numbers *in front* of the formulae of the substances in the equation.

Example 1.

Unbalanced

CH_4 $+$ O_2 $→$ CO_2 $+$ H_2O

Balanced

CH_4 $+$ $2O_2$ $→$ CO_2 $+$ $2H_2O$

Example 2.

Unbalanced

Mg $+$ HCl $→$ $MgCl_2$ $+$ H_2

Balanced

Mg $+$ $2HCl$ $→$ $MgCl_2$ $+$ H_2

Example 3.

Unbalanced

$AgNO_3$ $+$ $MgCl_2$ $→$ $AgCl$ $+$ $Mg(NO_3)_2$

Balanced

$2AgNO_3$ $+$ $MgCl_2$ $→$ $2AgCl$ $+$ $Mg(NO_3)_2$

5.5 State symbols

Chemical equations are very useful because they tell us exactly what happens when substances react. To give as much information as possible, we can indicate the physical states of the substances—whether they are solids, liquids or gases.

For example, the reaction between magnesium and oxygen would be written:

$2Mg(s) + O_2(g) → 2MgO(s)$

where (s) stands for solid and (g) for gas. The reaction between sodium and water is written:

$2Na(s) + 2H_2O(l) → 2NaOH(aq) + H_2(g)$

(l) stands for liquid and (aq) for a substance in aqueous solution.

The conditions of the reaction must be taken into account when writing state symbols.

For example, copper(II) oxide reacts with hydrogen according to the equation:

$$CuO + H_2 \rightarrow Cu + H_2O$$

This reaction only occurs on heating, so the water formed will be steam (a gas) and not a liquid. The equation is written:

$$CuO(s) + H_2(g) \rightarrow Cu(s) + H_2O(g)$$

5.6 Ionic equations

If aqueous sodium hydroxide is added to aqueous copper(II) sulphate a light blue precipitate is formed. This blue substance is copper(II) hydroxide.

sodium hydroxide + copper(II) sulphate →
copper(II) hydroxide + sodium sulphate

The chemical equation for this reaction is written:

$$2NaOH(aq) + CuSO_4(aq) \rightarrow$$
$$Cu(OH)_2(s) + Na_2SO_4(aq)$$

What has happened is that copper(II) ions (Cu^{2+}) have combined with hydroxide ions (OH^-) to form solid copper(II) hydroxide. The formula $Na_2SO_4(aq)$ means that we have an aqueous mixture of sodium ions (Na^+) and sulphate ions (SO_4^{2-}).

Sodium ions were present in the aqueous sodium hydroxide and sulphate ions in the aqueous copper(II) sulphate. These two ions have not been affected in the reaction. We can write the chemical equation showing all the ions present:

$$2Na^+OH^-(aq) + Cu^{2+}SO_4^{2-}(aq) \rightarrow$$
$$Cu^{2+}(OH^-)_2(s) + (Na^+)_2SO_4^{2-}(aq)$$

This can be simplified by leaving out all ions unaffected in the reaction. If we leave out the sodium ions and the sulphate ions we get:

$$2OH^-(aq) + Cu^{2+}(aq) \rightarrow Cu(OH)_2(s)$$

We have now written an *ionic equation*. This equation must be balanced in terms of the number of atoms and charges.

If aqueous sodium chloride is added to aqueous silver nitrate, a white precipitate is formed. This is silver chloride, because all sodium compounds are soluble in water. The reaction that has occurred is between silver ions (Ag^+) and chloride ions (Cl^-). The ionic equation for the reaction is:

$$Ag^+(aq) + Cl^-(aq) \rightarrow AgCl(s)$$

If hydrochloric acid is added to aqueous sodium carbonate, a gas is given off. The chemical equation for the reaction is:

$$Na_2CO_3(aq) + 2HCl(aq) \rightarrow$$
$$2NaCl(aq) + CO_2(g) + H_2O(l)$$

Sodium ions and chloride ions were present in solution in the reactants and are still present in solution in the products. They have not changed in the reaction. The reaction involves carbonate ions (CO_3^{2-}) and hydrogen ions (H^+), giving carbon dioxide and water. The ionic equation is:

$$CO_3^{2-}(aq) + 2H^+(aq) \rightarrow CO_2(g) + H_2O(l)$$

If aqueous potassium carbonate and dilute sulphuric acid had been used, the ionic equation would be the same. This is because potassium ions and sulphate ions play no part in the reaction.

When writing ionic equations remember: solutions of acids and alkalis contain ions. Salts are shown as ionic unless insoluble in water.

5.7 Reduction and oxidation—redox reactions

When magnesium burns in oxygen the following reaction occurs:

magnesium + oxygen → magnesium oxide

The balanced equation for this reaction is:

$$2Mg(s) + O_2(g) \rightarrow 2MgO(s)$$

Because magnesium has gained oxygen, we say that it has been oxidised and the process is known as *oxidation*.

Magnesium oxide is an ionic compound, so the magnesium atom has lost two electrons to become the magnesium ion Mg^{2+}. Another definition is that *oxidation is a loss of electrons*.

The oxidation number of magnesium is zero and the oxidation number of magnesium in magnesium oxide is +2. The oxidation number of magnesium has increased by 2. Another definition of oxidation is that *oxidation is an increase in oxidation number*.

Look at the reaction between copper(II) oxide and hydrogen:

copper(II) oxide + hydrogen → copper + steam
$$CuO(s) + H_2(g) \rightarrow Cu(s) + H_2O(g)$$

From these definitions, the hydrogen has been oxidised. It has gained oxygen, and its oxidation number has increased from zero to +1. *If oxidation occurs in a reaction, then reduction must also occur.*

The copper(II) oxide must have been reduced. *Reduction is either the loss of oxygen or a gain of electrons* or *a decrease in the oxidation number*.

There is another definition of reduction and that is the addition of hydrogen.

Similarly, oxidation can be defined as the removal of hydrogen.

Reactions in which reduction and oxidation occur are known as *redox reactions*.

Therefore *oxidation* is:
1. adding oxygen,
2. removing hydrogen,
3. removing electrons, or
4. increasing the oxidation number;

and *reduction* is:
1. removing oxygen,
2. adding on hydrogen,
3. adding electrons, or
4. lowering the oxidation number.

The definition in terms of electrons can be remembered by OIL RIG.

Oxidation Is Loss of electrons.
Reduction Is Gain of electrons.

In a REDOX reaction the substance which is reduced does the oxidising. It is known as the **oxidising agent.**

The substance which is oxidised does the reducing. It is known as the **reducing agent.**

In the reaction between hydrogen and oxygen to form water, the hydrogen is the reducing agent and the oxygen is the oxidising agent.

Table 3 shows some common oxidising and reducing agents.

OXIDISING AGENTS	REDUCING AGENTS
oxygen	hydrogen
nitric acid	carbon monoxide
concentrated sulphuric acid	sulphur dioxide
potassium manganate(VII)	

Table 3

Examples of Redox reactions

Look at the following reactions:

$$2H_2 + O_2 \rightarrow 2H_2O$$

The hydrogen has been oxidised, it has gained oxygen and its oxidation number has increased from 0 to +1.

The oxygen has been reduced, it has gained hydrogen and its oxidation number has decreased from 0 to −2.

$$2FeCl_2 + Cl_2 \rightarrow 2FeCl_3$$

The ionic equation for this reaction is

$$2Fe^{2+} + Cl_2 \rightarrow 2Fe^{3+} + 2Cl^-$$

The Fe^{2+} in $FeCl_2$ has been oxidised: it has lost an electron to become Fe^{3+}, and its oxidation number has increased from +2 to +3.

The chlorine molecule has been reduced. Each atom has gained an electron to form a Cl^- ion, and its oxidation number has decreased from 0 to −1.

$$3CuO + 2NH_3 \rightarrow 3Cu + N_2 + 3H_2O$$

The copper(II) oxide has been reduced. It has lost oxygen: each Cu^{2+} ion has gained electrons to form a Cu atom, and the oxidation number of copper has decreased from +2 to 0.

The ammonia has been oxidised. It has lost hydrogen to become nitrogen, and the oxidation number of nitrogen has increased from −3 to 0.

In this reaction, hydrogen and oxygen have not undergone any redox reaction. Their oxidation number remains at +1 for hydrogen and −2 for oxygen.

Tests for oxidising agents and reducing agents

The easiest way to test for an oxidising agent is to add it to a substance which is easily oxidised (ie a

reducing agent). In the same way, the easiest way to test for a reducing agent is to add it to a substance that is easily reduced (ie an oxidising agent).

Tests for oxidising agents:
1. Add the substance to aqueous potassium iodide, acidified with dilute sulphuric acid.

Result: The solution turns brown because iodine is formed. The presence of iodine can be confirmed by adding starch solution (a blue colour is formed), or by shaking with 1,1,1-trichloroethane.

2. Add the substance to freshly prepared iron(II) sulphate solution acidified with dilute sulphuric acid.

Result: The solution changes from pale green to yellow because iron(III) ions are formed. Iron(III) ions can be confirmed by adding sodium hydroxide solution. A reddish brown precipitate is formed.

Tests for reducing agents:
1. Add potassium manganate(VII) solution acidified with dilute sulphuric acid drop by drop to the substance.

Result: The purple colour of potassium manganate(VII) changes and becomes colourless.

2. Add aqueous potassium dichromate(VI) acidified with dilute sulphuric acid drop by drop to the substance.

Result: The orange colour of the potassium dichromate(VI) changes to green.

3. Add a few drops of aqueous bromine to the substance.

Result: The reddish brown colour of bromine changes and becomes colourless.

Experiment 5.1 To test oxidising agents and reducing agents

You will need solutions of the following substances: sodium sulphite, tin(II) chloride, hydrogen peroxide, iron(III) chloride.

Carry out the tests for oxidising agents and reducing agents on each of these four substances.

Do any of the substances behave as both oxidising agents and reducing agents?

Questions

1. How many elements are there in the compound methane (CH_4)?

A 1   D 4
B 2   E 5
C 3

2. Which one of the following equations correctly represents the combustion of methane (CH_4) in an excess of oxygen?

A $CH_4(g) + O_2(g) \rightarrow C(s) + 2H_2O(g)$
B $CH_4(g) + O_2(g) \rightarrow CO_2(g) + 2H_2(g)$
C $CH_4(g) + 2O_2(g) \rightarrow CO_2(g) + 2H_2O(g)$
D $2CH_4(g) + O_2(g) \rightarrow 2CO(g) + 4H_2(g)$
E $2CH_4(g) + 3O_2(g) \rightarrow 2CO(g) + 4H_2O(g)$

3. When iron(II) sulphate ($FeSO_4$) reacts with silver nitrate ($AgNO_3$), a reaction occurs. A precipitate of silver is formed. Which one of the following is the ionic equation for this reaction?

A $Fe^{2+}(aq) + 2Ag(s) \rightarrow Fe(s) + 2Ag^+(aq)$
B $Fe^{3+}(aq) + 3Ag(s) \rightarrow Fe(s) + 3Ag^+(aq)$
C $Fe^{3+}(aq) + Ag(s) \rightarrow Fe^{2+}(aq) + Ag^+(aq)$
D $Fe^{2+}(aq) + Ag^+(aq) \rightarrow Fe^{3+}(aq) + Ag(s)$
E $Fe(s) + 2Ag^+(aq) \rightarrow Fe^{2+}(aq) + 2Ag(s)$

4. In which one of the following pairs of chemicals does the named element show the same oxidation state (number)?

A copper in Cu_2O and CuO
B iron in FeO and Fe_2O_3
C manganese in MnO_2 and $KMnO_4$
D oxygen in O_2 and MgO
E sulphur in SO_3 and H_2SO_4

5. In which one of the following reactions does the oxidation state (number) of chlorine increase by one:

A $Cl_2(g) + H_2(g) \rightarrow 2HCl(g)$
B $3Cl_2(g) + 6NaOH(aq) \rightarrow 5NaCl(aq) + NaClO_3(aq) + 3H_2O(l)$
C $NaCl(aq) + AgNO_3(aq) \rightarrow AgCl(s) + NaNO_3(aq)$
D $MnO_2(s) + 4HCl(aq) \rightarrow MnCl_2(aq) + Cl_2(g) + 2H_2O(l)$
E $2KClO_3(s) \rightarrow 2KCl(s) + 3O_2(g)$

6. How many atoms are there in each of the following formula units?
a SO_2 b Pb_3O_4 c $CuSO_4$
d N_2 e ZnO f $(NH_4)_2SO_4$
g CH_4 h NO_2 i CO
j C_4H_{10}

7. Write the names of the compounds which have the following formulae:
a $MgCl_2$ b $AgNO_3$ c $ZnSO_4$
d Al_2O_3 e $CaCO_3$ f NH_4Cl
g $NaOH$ h $CuSO_4$ i FeS
j Na_2SO_3

8. Balance the following equations:
a $Mg + O_2 \rightarrow MgO$
b $CO + O_2 \rightarrow CO_2$
c $Fe_2O_3 + H_2 \rightarrow Fe + H_2O$
d $Na_2CO_3 + HCl \rightarrow NaCl + H_2O + CO_2$
e $FeCl_2 + Cl_2 \rightarrow FeCl_3$
f $Na + H_2O \rightarrow NaOH + H_2$
g $AgNO_3 + CaCl_2 \rightarrow AgCl + Ca(NO_3)_2$
h $NaNO_3 \rightarrow NaNO_2 + O_2$
i $C_2H_6 + O_2 \rightarrow CO_2 + H_2O$
j $H_2 + Cl_2 \rightarrow HCl$

9. $NaNO_3$ NH_3 $CuSO_4$ $MgCO_3$ Al_2O_3
From the list of compounds shown by these formulae, select one compound which contains:

a an element which is in Group II of the Periodic Table
b a metallic element showing an oxidation number of +3
c a non-metallic element showing an oxidation number of -3
d a non-metallic element showing an oxidation number of +5

[C]

10. Write balanced equations for the following reactions:

a zinc + oxygen(O_2) $\rightarrow$ zinc oxide
b lead nitrate + sodium chloride $\rightarrow$ lead chloride + sodium nitrate
c magnesium + sulphuric acid(H_2SO_4) $\rightarrow$ magnesium + hydrogen(H_2)
sulphate
d copper(II) + sulphuric $\rightarrow$ copper(II) + carbon(CO_2) + water
carbonate   acid   sulphate   dioxide
e copper(II) + sodium $\rightarrow$ copper(II) + sodium
sulphate   hydroxide   hydroxide   sulphate
f zinc + hydrochloric $\rightarrow$ zinc + water
oxide   acid(HCl)   chloride
g iron + sulphur $\rightarrow$ iron(II) sulphide
h zinc + copper(II) chloride $\rightarrow$ zinc chloride + copper
i ammonium + calcium $\rightarrow$ calcium + ammonia(NH_3) + water
chloride   hydroxide   chloride
j calcium + carbon $\rightarrow$ calcium + water
hydroxide   dioxide   carbonate

11. Give the oxidation state of nitrogen in each of the following:
(i) NH_3 (ii) NO (iii) NO_2 (iv) NO_3^- [C]

12. Make use of the information in the table below to answer questions **a.** to **f.**

Name	Formula	Melting point/°C	Boiling point/°C	Behaviour with water
Magnesium chloride	$MgCl_2$	714	1418	Dissolves—no apparent reaction
Phosphorus trichloride	PCl_3	−92	76	Reacts giving acidic liquid
Potassium chloride	KCl	790	1407	Dissolves—no apparent reaction
Silicon tetrachloride	$SiCl_4$	−70	58	Reacts giving acidic liquid
Strontium chloride	$SrCl_2$	875	1250	Dissolves—no apparent reaction
Disulphur dichloride	S_2Cl_2	−80	138	Reacts giving acidic liquid

a Give *one general* physical difference between the metallic chlorides and the non-metallic chlorides.

b Give *one general* chemical difference between the metallic chlorides and the non-metallic chlorides.

c In which of the chlorides in the table are the bonds formed by the sharing of electrons?

d Assuming that silicon and strontium do not show variable valency, write the formulae of their oxides. Give *one* chemical difference you would expect between these oxides.

e The reaction between silicon tetrachloride and water can be represented by the equation

$SiCl_4 + 4H_2O \rightarrow H_4SiO_4 + 4HCl$

Write an equation for the reaction you would expect between phosphorus trichloride and water.

f When sulphur dichloride reacts with water, it gives a pale yellow precipitate and a solution that contains hydrochloric acid and sulphurous acid (H_2SO_3). State what you think the precipitate is. Write an equation for the reaction between disulphur dichloride and water.

[C]

6 Sorting out the elements—The Periodic Table

Chemists in the 19th century had a problem. Over sixty elements had been discovered. Many of their compounds had been prepared. There was a mountain of information, but this was not organised. The elements had to be grouped in some way so that similarities between elements could be noted; patterns and trends could be observed. Only if chemists managed to organise their facts could the study of chemistry advance.

6.1 Metals and non-metals

If you were asked to divide elements into groups, what groups would you choose? You might divide them into solids, liquids and gases. How useful would this be? One of the earliest attempts to group the elements was to divide them into metals and non-metals.

Many metals have physical and chemical properties in common. The same applies to non-metals. Physical properties are properties such as melting point, density and appearance. They do not involve any chemical reactions. The chemical properties of a substance are the chemical reactions of that substance.

Table 1 shows how the properties of metals and non-metals differ.

This is a useful start at grouping the information. If you were told that a new element was a metal you would know a lot about it before you even saw a lump of it.

Unfortunately the metal/non-metal grouping is not completely reliable, as there are some exceptions. Can you think of a metal element that melts below room temperature? Can you think of a non-metal element that is a good conductor of electricity? Can you think of a metal element that is less dense than water? These elements do exist. You can use the data page at the back of the book (page 138) to identify them.

The greatest disadvantage of the metal/non-metal grouping is that it only divides the elements into two groups. Chemists had to find a way of dividing the metals and non-metals into smaller groups.

6.2 The Periodic Table of the elements

The modern Periodic Table greatly improves on the metal/non-metal grouping of the elements, In the Periodic Table, elements are arranged in order of increasing atomic number. Elements with similar chemical properties are found in the same vertical columns.

In Chapter 3 we saw the first twenty elements listed in order of increasing atomic number. We can get some idea about how the Periodic Table was obtained by looking at these elements again.

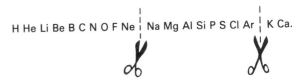

H He Li Be B C N O F Ne ¦ Na Mg Al Si P S Cl Ar ¦ K Ca.

Suppose we cut the list every time we come to an element similar to one of lower atomic number. We might cut at the positions shown because lithium (Li), sodium (Na) and potassium (K) are very similar elements. If we take the strips formed by cutting the list of elements, we can arrange them so that similar elements are in vertical columns. We get

```
H  He Li Be  B  C  N  O  F  Ne
         Na Mg Al Si  P  S  Cl Ar
         K  Ca
```

By this method the first 20 elements are arranged as they are in the modern Periodic Table.

The first thing to realise about the Periodic Table is that it divides the elements into metals and non-metals. You can see this by looking at the

Table 1

PROPERTIES OF METALS	PROPERTIES OF NON-METALS
Physical properties	
1. Can be bent and hammered without breaking	**1.** Are brittle (easily broken) when solid
2. Are good conductors of electricity	**2.** Are poor conductors of electricity
3. Are good conductors of heat	**3.** Are poor conductors of heat
4. Usually have a high melting point	**4.** Usually have a low melting point
5. Usually have high densities	**5.** Usually have low densities
Chemical properties	
6. React with oxygen to form basic oxides. (These neutralise acids)	**6.** React with oxygen to form acidic oxides. (These neutralise alkalis)

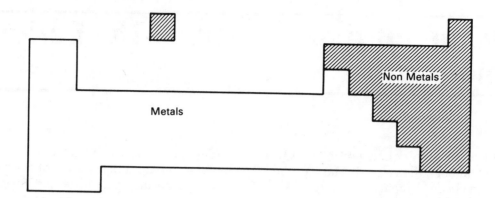

Fig 6.1 The Periodic Table

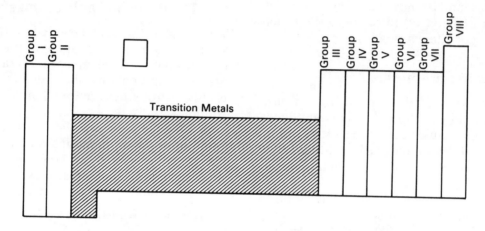

Fig 6.2 The groups of the Periodic Table

full Periodic Table on page 35 or by looking at the outline of the Periodic Table in Fig 6.1.

The elements to the left of the 'steps' are metals. Those to the right are non-metals. There are many more metals than non-metals.

The Periodic Table also divides the elements up into horizontal rows and vertical columns. *The horizontal rows are called Periods. The vertical columns are called Groups.*

Let us now look at some Groups to see how similar the elements in them are.

Group I—the alkali metals

Li	lithium
Na	sodium
K	potassium
Rb	rubidium
Cs	caesium
Fr	francium

The elements in Group I of the Periodic Table are known as the alkali metals. You may have seen samples of lithium, sodium and potassium in the laboratory. You are very unlikely to see samples of rubidium, caesium or francium. However, because the elements in a Group are similar, it should be possible to predict the properties of rubidium, caesium and francium.

The alkali metals have the following properties in common:

a they are soft metals that can be cut with a knife,
b they have low densities,
c they are stored under oil because they react with air and water,
d they form alkaline solutions and hydrogen when they react with water,
e all alkali metal compounds dissolve in water.
f The compounds of the alkali metals have similar formulae:

LiCl
NaCl
KCl
RbCl
CsCl
FrCl

The formulae are similar because the alkali metals all have the same number of valency electrons (outer shell electrons). The Group I elements all have one outer shell electron.

element	electron arrangement
Li	2,1
Na	2,8,1
K	2,8,8,1

(The number of outer shell electrons is always the same as the Group number.)

The Periodic Table

Group I	II											III	IV	V	VI	VII	VIII
				1 H 1													4 He 2
7 Li 3	9 Be 4											11 B 5	12 C 6	14 N 7	16 O 8	19 F 9	20 Ne 10
23 Na 11	24 Mg 12											27 Al 13	28 Si 14	31 P 15	32 S 16	35.5 Cl 17	40 Ar 18
39 K 19	40 Ca 20	45 Sc 21	48 Ti 22	51 V 23	52 Cr 24	55 Mn 25	56 Fe 26	59 Co 27	59 Ni 28	64 Cu 29	65 Zn 30	70 Ga 31	73 Ge 32	75 As 33	79 Se 34	80 Br 35	84 Kr 36
85 Rb 37	88 Sr 38	89 Y 39	91 Zr 40	93 Nb 41	96 Mo 42	– Tc 43	101 Ru 44	103 Rh 45	106 Pd 46	108 Ag 47	112 Cd 48	115 In 49	119 Sn 50	122 Sb 51	128 Te 52	127 I 53	131 Xe 54
133 Cs 55	137 Ba 56	La to Lu	178 Hf 72	181 Ta 73	184 W 74	186 Re 75	190 Os 76	192 Ir 77	195 Pt 78	197 Au 79	201 Hg 80	204 Tl 81	207 Pb 82	209 Bi 83	– Po 84	– At 85	– Rn 86
– Fr 87	– Ra 88																
		– Ac 89	– Th 90	– Pa 91	– U 92												

In each box: a X b

a = relative atomic mass
X = atomic symbol
b = atomic number

Fig 6.3

The Group I metals are similar, but they are not identical.

This can be seen by reacting lithium, sodium and potassium with water. All three react to form an alkaline solution and hydrogen gas is produced. However, they do not all react at the same rate. Lithium reacts fairly slowly. Sodium reacts faster and enough heat is produced to melt the sodium. Potassium reacts even faster. Enough heat is produced to melt the potassium and to set fire to the hydrogen gas as it is formed.

lithium + water → lithium hydroxide + hydrogen

$$2Li(s) + 2H_2O(l) \rightarrow 2LiOH(aq) + H_2(g)$$

similarly

$$2Na(s) + 2H_2O(l) \rightarrow 2NaOH(aq) + H_2(g)$$

$$2K(s) + 2H_2O(l) \rightarrow 2KOH(aq) + H_2(g)$$

You should see from this evidence that the Group I metals become more reactive going down the group. They become more reactive with increasing atomic number. The element francium, at the bottom of the group, must be very, very reactive.

Perhaps now you realise why you are unlikely to see a sample of francium. If you were given a sample of francium, you would keep it well away from water!

Group VII—The halogens

symbol for atom	*name*	*appearance*
F	fluorine	very pale yellow gas
Cl	chlorine	yellow-green gas
Br	bromine	red-brown liquid
I	iodine	purple-black solid
At	astatine	?

This group of elements is non-metallic. Their appearance shows a trend, in that the colours become darker as the atomic number increases.

Because astatine atoms are very unstable (they are radioactive), nobody has actually seen this element. We would expect the element to be a solid and that its colour would be black.

Unlike the Group I elements, when the halogens react with water they produce acidic solutions.

The compounds of the halogens have similar formulae and properties, for example,

HF, HCl, HBr, HI } acidic gases

The reactivity of the halogens increases going up the Group, so that fluorine is the most reactive of these elements.

Group VIII—The noble gases

The noble gases are helium, neon, argon, xenon, krypton and radon. They are found in Group VIII (sometimes called Group 0) of the Periodic Table.

The noble gases are very unreactive and very few compounds of the noble gases have ever been prepared. The noble gases are the only elements that exist as single atoms. This means that their atoms must be very stable.

Since the chemical properties of an element depend on the arrangement of electrons in its atoms, the noble gases must have a very stable electron arrangement. The properties of the noble gases have greatly helped chemists to understand how atoms of other elements react together to form compounds.

The Transition Metals

These have the following properties in common:

a they are strong, high melting point metals,
b they are dense metals,
c they form coloured (non white) compounds,
d they form insoluble oxides, hydroxides and carbonates,
e they or their compounds make good catalysts (see Chapter 7),
f they can show a number of valency states (oxidation states).

Suppose that a chemist was told that an element was a transition metal. He would know that it had the general properties of a *metal* and the special properties of a *transition metal*. He would have learnt a great deal about the element very easily.

Suppose you were given a metal carbonate and asked to find out what metal it contained. If it dissolved in water you would know that it could not be a transition metal carbonate. They are all insoluble. This simple experiment would eliminate a large number of possible metals. It would make your problem much easier.

Iron is a typical transition element and some of its chemistry is described in Chapter 11.

Periods

A large number of trends and patterns in the physical and chemical properties of elements can be detected as we move across a period. Use the data page to plot a graph of density of element against its atomic number. Do this for the first twenty elements. Is there any pattern in the way the densities of elements vary across a period?

As we move from left to right along any period there is a gradual change from metal to non-metal.

The formulae of compounds show a simple trend as we move across a period, for example,

$LiCl$ $BeCl_2$ BCl_3 CCl_4 NCl_3 OCl_2 FCl
$NaCl$ $MgCl_2$ $AlCl_3$ $SiCl_4$ PCl_3 SCl_2 $ClCl(Cl_2)$

Because of the change from metal to non-metal across a period, there is a change in the chemical

properties of the oxides of the elements (see page 33).

Na_2O MgO Al_2O_3 SiO_2 P_2O_5 SO_3 Cl_2O_7

basic   amphoteric   acidic

The oxidation number of the element combined with oxygen in these oxides corresponds to the Group in which that element occurs in the Periodic Table.

The Periodic Table allows similar elements to be grouped. It allows patterns and trends to be seen. It makes chemistry a much easier subject to study, because it organises the facts. As you work through this book you should often look at the Periodic Table. In this way you will realise many of the patterns and trends that exist.

Questions

1. P, Q and R are in the same period of the Periodic Table. The oxide of P dissolves in water to form a solution with a pH less than 7. The oxide of Q forms an oxide which dissolves in water to form a solution with a pH greater than 7. The oxide of R is soluble in both hydrochloric acid and in aqueous sodium hydroxide. The order of increasing atomic number for P, Q and R is:

A P, Q, R
B P, R, Q
C Q, P, R
D Q, R, P
E R, P, Q

2. Astatine is in Group VII of the Periodic Table. Its chemistry will resemble that of:

A carbon
B iodine
C nitrogen
D neon
E sodium.

3. In which group of the Periodic Table is the element with an atomic number of 5?

A 1
B 2
C 3
D 4
E 5

4. Which one of the following properties suggests that manganese is a transition metal?

A Manganese reacts with dilute hydrochloric acid to give hydrogen.
B Manganese has only one common isotope.
C Manganese compounds are coloured.
D Manganese burns in oxygen to form the oxide.
E Manganese reacts with steam at red heat.

5. Which one of the following statements about the Periodic Table is correct?

A All elements in the same Group of the Periodic Table have the same number of electrons.

B When Group I metals react with water they form acidic solutions.

C The noble gases are diatomic (two atoms per molecule).

D The carbonates of the transitions metals are soluble in water.

E The melting points of the elements in Group VII increase down the group.

6. Use a copy of the Periodic Table to identify the following elements from the information provided.

Element A It has an atomic number of 16.
Element B It has a relative atomic mass of 64.
Element C It has an electron arrangement 2,8,1.
Element D It is the least dense element in Group V.
Element E Atoms of element E contain 18 electrons.
Element F It has the symbol W.

7. The diagram below shows an outline of the Periodic Table with ten elements in position. You are to use the letters given for each element to answer the following.

Example: Name one metal element. Answer R

a Name two elements in the same group.
b Name two elements in the same period.
c Which element is in Group I of the Periodic Table?
d Name two transition metals.
e Name two noble gases.
f Name two non metal elements.
g Name an element in Group VII of the Periodic Table.
h Which element has the highest atomic number?
i Which element would be stored under oil?
j Name an element that exists as individual atoms?
k Name an element with a coloured (non-white) oxide.

8. A new element Devonium has been discovered. It is a soft element which is easily cut with a knife. It reacts violently with water forming hydrogen gas and leaving an alkaline solution. It is a very good conductor of electricity.

a Is Devonium a metal or a non metal? Give a reason for your answer.
b How would you store a sample of Devonium?
c Name an element similar to Devonium.
d In which group of the Periodic Table would you expect to find Devonium?
e How many electrons would you expect there to be in the outermost shell of a Devonium atom?
f Would you be surprised if you found a lump of Devonium on a river bank? Explain your answer

9. Describe the differences between metals and non-metals by referring to:

(i) their physical appearances
(ii) their electrical properties
(iii) the chemical properties of their oxides
(iv) the type of bond present in their chlorides.

(You may find it helpful to give your answers in the form of a table.)

[C]

10. Study the copy of the Periodic Table provided and answer the following questions:

a Explain, giving one example in each case, what you understand by:
(i) a group (ii) a period (iii) a transition metal.

b What determines the position of an element in the Periodic Table?

c Note the position of indium, atomic number 49, in the Table.
(i) Would you expect indium to be a metal or a non-metal?
(ii) Give the formula of the compound formed between indium and chlorine.
(iii) Give the formula of the compound formed between indium and oxygen.
(iv) Would you expect the oxide in (iii) to be acidic, basic or amphoteric?

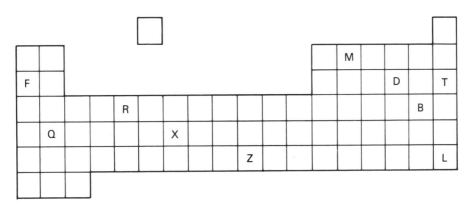

d Construct a table listing four differences in the properties of the metals as compared with the non-metals. Indicate which one of the four properties you consider to be the most reliable in classifying an element as a metal or a non-metal.

[C]

11. Use the information in the following passage to answer the questions on the element selenium which follow.

Selenium (Se) is an element in Group VI of the Periodic Table and it has a relative atomic mass of 79. The chemistry of selenium closely resembles that of sulphur. Selenium burns in oxygen to form an oxide which contains 71·2% by mass of selenium. When this oxide is heated with an excess of magnesium metal, a mixture of magnesium oxide and magnesium selenide is formed. Addition of dilute hydrochloric acid to this mixture results in the formation of the gas hydrogen selenide.

a How many valency (outer shell) electrons are there in one atom of selenium?

b Suggest: (i) the formulae for hydrogen selenide and magnesium selenide   (ii) the equation for the reaction between magnesium selenide and hydrochloric acid.

c (i) Calculate the formula of the oxide of selenium described above. Write the equation for the formation of this oxide. (Relative atomic mass: O, 16.)
   (ii) Write the equation for the reaction between this oxide of selenium and magnesium.

d Explain why the hydrogen selenide prepared by the method given in the passage is usually contaminated with hydrogen.

e What would you expect to observe if hydrogen selenide were passed into lead(II) nitrate solution? Give a reason for your answer.

[C]

7 Chemical reactions

7.1 Energy changes in reactions

Heat energy

We use energy to provide heat, air conditioning and light for our homes, offices, shops and factories. We use energy to power our cars, lorries, trains, boats and planes. We use energy in our factories to make the things we need, and the things that just make life more comfortable. We use an enormous amount of energy. Most of our energy is obtained from chemicals known as *fuels*. The fossil fuels are coal, oil and gas. They were formed by nature millions of years ago.

The formation of coal

Coal was formed about 200–300 million years ago. Plants in great forests died and fell to the ground. Over a long period of time layers of mud, sand and rock formed over the layers of plant matter. They were pushed down deeper and deeper into the Earth's crust. Heat and pressure gradually changed the plant matter into coal. The deeper the coal was pushed the more it was compressed into harder and harder coal.

The formation of oil and gas

Oil and gas were also formed many millions of years ago. Shallow coastal seas existed at that time, that were rich in animal life. As tiny creatures in these seas died they sank to the bottom and started to decay. Gradually rivers washed mud, sand and rock down into the seas and the deposits of dead creatures were buried. They were changed by heat and pressure to produce oil and gas. As layers of rock formed on top of the oil and gas, it was squeezed and forced through any porous rock. Only when it came in contact with a hard non-porous rock was the oil and gas trapped.

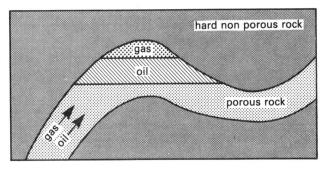

Fig. 7.1

Exothermic and endothermic processes

The burning of fuels involves a reaction with oxygen. For coal, which is mainly carbon, the reaction can be written as:

$$C + O_2 \rightarrow CO_2$$

For natural gas, the reaction can be written as:

$$CH_4 + 2O_2 \rightarrow CO_2 + 2H_2O$$

Both processes *give out* large amounts of heat, and they are called *exothermic* processes.

If you dissolve ammonium chloride in water you will see that the temperature of the mixture drops as the process of dissolving occurs.

This is the opposite of an exothermic process because heat energy is being *absorbed*. This type of process is called an *endothermic process*.

Experiment 7.1 Exothermic and endothermic reactions

1. Place a 3 cm depth of dilute hydrochloric acid in a boiling tube and record its temperature. Add a 2 cm length of magnesium ribbon and stir the mixture carefully with the thermometer.

What happens to the temperature of the mixture? Is the reaction exothermic or endothermic?

2. Place 25 cm³ of dilute hydrochloric acid in a beaker and measure its temperature. Add about 25 cm³ of a saturated solution of potassium hydrogencarbonate.

What happens to the temperature of the solution? Is this reaction exothermic or endothermic?

3. Place a 3 cm depth of dilute hydrochloric acid in a boiling tube and measure its temperature. Add an approximately equal volume of dilute sodium hydroxide. Is the reaction exothermic or endothermic?

Forming chemical bonds is an exothermic process and breaking bonds is an endothermic process. Whether a chemical reaction gives out or absorbs heat depends on the balance between these two processes.

Some chemistry books contain equations like those shown below where the symbol ΔH appears. This symbol means 'heat change'—if it has a *negative* value then the reaction is *exothermic*, if it is *positive* the reaction is *endothermic*. The values are quoted in kilojoules (kJ) which are units of heat.

1. $N_2 + 3H_2 \rightleftharpoons 2NH_3$ $\Delta H = -92\,\text{kJ/mol}$
2. $2SO_2 + O_2 \rightleftharpoons 2SO_3$ $\Delta H = -187\,\text{kJ/mol}$
3. $H_2 + I_2 \rightleftharpoons 2HI$ $\Delta H = +50\,\text{kJ/mol}$

From this information, we can tell that **1** and **2** are exothermic. The formation of hydrogen iodide shown in reaction **3** is clearly an endothermic process.

Energy from the nucleus

For about 80 years it has been known that some elements such as uranium are unstable. Atoms of these elements spontaneously break down to form atoms of different elements. These unstable elements are known as *radioactive* elements. The spontaneous breaking down of their nuclei is known as *radioactivity*.

When radioactive elements break down large quantities of energy are released. This energy is used in nuclear power stations to produce electricity.

Light energy

When coal and natural gas burn, you see a flame. This is light energy given out in the reaction.

In other processes, light energy is absorbed. For example, in photosynthesis:

water + carbon dioxide + sunlight →

glucose + oxygen

Hydrogen and chlorine do not react in the dark, but when exposed to light a reaction rapidly occurs:

$$H_2 + Cl_2 + light \rightarrow 2HCl$$

Absorption of light energy is important in photography. The film in a camera is coated with a silver compound (silver bromide or silver iodide). When the shutter is opened, light causes a chemical reaction to occur and silver is formed. The silver in the compound has been reduced to silver metal. These metal particles cause dark patches on the film, thus producing a negative.

Electrical energy

Electrical energy can be produced by chemical reactions. If the apparatus is set up as shown, the lamp glows. This shows that an electric current is produced. Copper and magnesium differ widely in reactivity and this makes the bulb glow brightly. If different metals are used, the more similar their reactivity the dimmer the glow.

Similar chemical reactions occur in the batteries in portable transistor radios. Batteries are convenient portable energy sources. A car battery is not as easy to carry as a radio battery, but it does produce a lot more energy!

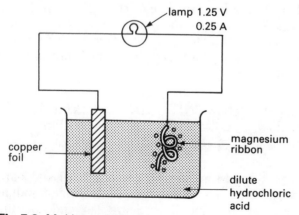

Fig 7.2 Making a battery

We can use electricity to make reactions occur. This process is called *electrolysis* and is described in Chapter 10.

7.2 Speed of reaction

The chemical changes that take place when a fruit ripens are very slow. We do not blink and find that a fruit has ripened; blink again and find it overripe. In fact the changes are so slow that we cannot see them happening. On the other hand, if a match is put to a mixture of natural gas and oxygen it might explode. The reaction may be complete in a fraction of a second; so fast that we cannot see it happening. Chemical reactions vary from very fast to very slow.

This can give problems to chemists in industry. They obviously do not want reactions to take place at an explosive rate, but they do want them to take place fast enough to make the best use of their chemical plant. Chemists want to be able to control the speed of chemical reactions. They may want to be able to speed them up or slow them down.

Before chemists could hope to *control* the speed of chemical reactions they had to find out what factors affect the speeds of reactions. We have already seen how light can affect reactions, other factors are:

1. particle size,
2. concentration,
3. temperature,
4. the presence of a catalyst,
5. the type of bonding present in the reactants.

1. The size of particles

When a solid reacts with a liquid or a gas, the size of the solid particles affects the speed of the reaction.

The smaller the solid particles, the faster the rate of the reaction.

This can be shown in the laboratory by the following simple experiment.

Experiment 7.2 The effect of particle size on the speed of a reaction

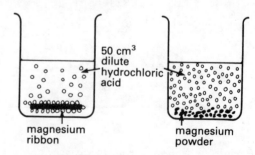

Fig 7.3 The reaction of magnesium and acid

Using a measuring cylinder carefully place 50 cm³ of dilute hydrochloric acid into each of two small beakers. Take a 10 cm length of magnesium ribbon and weigh it. Then weigh an equal mass of magnesium powder. Drop the piece of magnesium ribbon into one beaker of acid, starting a stopwatch as you do so. Note how long it takes for the magnesium to completely react. Repeat the experiment by putting the magnesium powder in the second beaker of acid.

You should find that the powder reacts in a shorter time. It must therefore have reacted faster. Would you expect the magnesium ribbon to react faster or slower if it had been rolled up into a tight reel?

Explanation

You should see from Fig 7.4 that if a block of magnesium is placed in acid, the acid can only react with the magnesium on the surface of the block. The speed of the reaction depends on the surface area of the block.

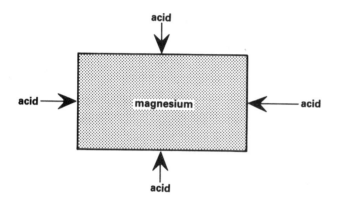

Fig 7.4

If the block of magnesium is cut into smaller pieces extra surfaces are made available for the acid to attack. This is shown in Fig 7.5.

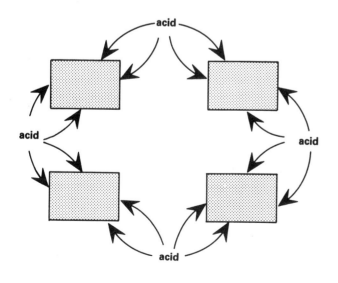

Fig 7.5

The acid can react with more magnesium at any one time, so the speed of the reaction increases. The more the magnesium is broken down into smaller particles, the larger its surface area becomes and therefore the faster it reacts.

Experiment 7.3 The effect of surface area on the reaction between marble chips and hydrochloric acid

This experiment is best carried out on a top pan balance.

Place 6 g of marble chips in a conical flask on a top pan balance. Add 40 cm³ of 2 mol/dm³ hydrochloric acid, place a loose plug of cotton wool in the neck of the flask. Immediately note the mass and start a clock. Note the mass every 30 seconds for ten minutes.

Plot a graph of mass on the vertical axis against time on the horizontal axis.

Repeat the experiment but crush the 6 g of marble chips so that the pieces are much smaller. Plot a graph of the results on the same axes as before.

Compare your results with the graphs shown below.

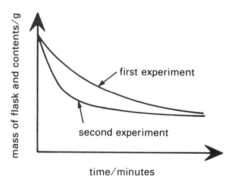

Fig 7.6

What is the purpose of the cotton wool plug in this experiment?

Why does the mass in each experiment decrease?

Why does each curve eventually become a horizontal line?

At what time is the slope of each curve the greatest?

In which experiment is the initial rate of reaction the greater?

It is worth realising that some solids, which we think of as harmless can be very dangerous as fine powders. We store coal in sheds or bunkers near our homes. We do not think of bags of flour as a fire hazard in our kitchens. However, coal dust and finely powdered flour can burn explosively in air. Great care has to be taken in coal mines and in flour mills to control dust and sparks. Insufficient control has led to a number of disasters.

2. Concentration

The concentration of reacting substances is a second factor that affects the speed of chemical reactions.

If the concentration of a reagent increases, the rate at which it reacts increases.

This can once again be shown in the laboratory by using the reaction between magnesium and hydrochloric acid.

Experiment 7.4 The effect of concentration on the reaction between magnesium and dilute hydrochloric acid

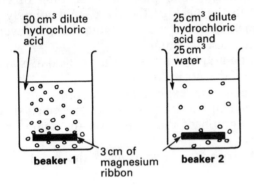

Fig 7.7

Place 50 cm³ of dilute hydrochloric acid into a small beaker. Into a second small beaker place 25 cm³ of dilute hydrochloric acid and 25 cm³ of water. Both beakers now contain 50 cm³ of acid solution, but the acid will be twice as concentrated in the first beaker. Cut two pieces of magnesium ribbon exactly 3 cm long. Drop one piece of magnesium into the first beaker starting a stopwatch as you do so. Note the time it takes for the magnesium to completely react. Repeat the experiment by dropping the other piece of magnesium into the second beaker.

You will find that the magnesium completely reacts in a shorter time in the more concentrated acid. This means that it reacts faster in the more concentrated acid.

Explanation

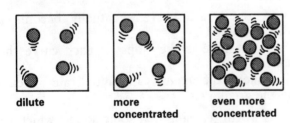

Fig 7.8

As a solution containing a substance becomes more concentrated there are more particles of that substance in a given space. This is shown in Fig 7.8. The greater the concentration of particles the greater the chance they have of colliding with each other.

If particles collide more often they will react more often. Therefore increasing the concentration of a substance increases the rate at which it will react.

Experiment 7.5 A further look at the reaction between magnesium and dilute hydrochloric acid

We can study the reaction between magnesium and an acid more completely using the apparatus shown.

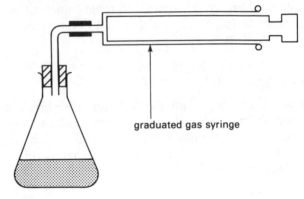

graduated gas syringe

Fig 7.9

Clamp a 100 cm³ gas syringe securely in a horizontal position. Connect the tubing and bung to the syringe.

Pour 50 cm³ (an excess) of 0·5 mol/dm³ hydrochloric acid into the flask using a funnel. Take a 9 cm length of clean magnesium ribbon and wind it into a loose coil. Hold the flask at such an angle that you can balance the magnesium in the neck of the flask.

Carefully insert the bung and then allow the magnesium to drop into the acid, starting a clock at the same time.

Read the time on the clock as the syringe reads 10, 20, 30 cm³ etc. Note the final reading on the syringe. You should give the piston of the syringe an occasional twist to keep it moving freely.

Repeat the whole experiment, using the same length of magnesium ribbon but with acid twice the concentration used in the first experiment.

Using the same axes, plot your results with syringe reading on the vertical and time on the horizontal. Compare your graphs with those shown below:

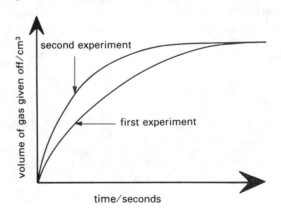

Fig 7.10

Look at your graphs and think about the following questions and answers.

Q. Why do both experiments produce about the same amount of hydrogen although the amount of acid present is very different?

A. The actual amount of acid used does not affect the volume of gas produced so long as the acid is in excess. The amount of magnesium used was the same in both experiments.

Q. Why does the curve in each case finish up as a horizontal line?

A. The horizontal line shows that no more gas is being given off. The reaction has stopped because all the magnesium has been used up. (There is still plenty of acid left.)

Q. At what time is the speed of reaction the greatest?

A. It should normally be at the start of the reaction (zero time).

This is because the acid is at its most concentrated at this time. As the reaction proceeds, acid is gradually used up and becomes more dilute. We would expect the graph to show its steepest slope at the beginning.

However, our experimental error is probably quite large at the beginning and also any oxide film on the metal surface will have to react first before any hydrogen is produced.

For reactions between gases, an increase in concentration occurs if the pressure is increased.

A rise in pressure squeezes the same number of particles into a smaller volume. The gas molecules will have a greater chance of colliding and the rate of reaction will increase.

3. Temperature

Temperature can have a very large effect on the speed of a chemical reaction.

If the temperature of the reactants is increased, they will react faster.

For many reactions increasing the temperature by 10 °C makes the reaction take place about twice as fast.

We can show the effect of temperature on the rate of a reaction in the laboratory by using the reaction between sodium thiosulphate solution and dilute nitric acid. When dilute nitric acid is added to sodium thiosulphate solution, a fine deposit of sulphur is formed. This makes the solution cloudy. As more and more sulphur is formed the solution becomes more and more cloudy. Eventually it is impossible to see through the solution.

Experiment 7.6 The effect of temperature on the reaction between sodium thiosulphate and nitric acid

By using a measuring cylinder, place 50 cm^3 of distilled water in a beaker. Label two clean test tubes A and B.

Fill the tube labelled A to the brim with 0·2 mol/dm^3 sodium thiosulphate, then pour the contents into the beaker of water. Heat the solution to between 45 °C and 50 °C, then place the beaker on a piece of paper marked with a cross.

Fill test tube B to the brim with 1 mol/dm^3 nitric acid. Note the temperature of the solution in the beaker. Add the contents of tube B, starting a clock as you do so. Swirl the mixture once and note how long it takes for the cross to disappear looking down through the solution in the beaker.

Repeat the experiment four more times. Each time your solution should be 5 °C cooler at the start.

Plot a graph of temperature against the time taken for the cross to disappear.

How long would it have taken for the cross to disappear if the experiment had been carried out at (i) 40 °C, (ii) 30 °C?

The cooking of food involves chemical reactions. The speed at which food is cooked by boiling is normally fixed because the temperature is always 100 °C, as that is the boiling point of water at atmospheric pressure. The boiling point of a liquid increases as the pressure above it increases. The cooking process can be speeded up considerably by using a pressure cooker. This cooking vessel allows the pressure inside to become greater than atmospheric pressure and therefore the cooking is speeded up because the water is boiling at a temperature greater than 100 °C.

4. Catalysts

Catalysts are substances that speed up chemical reactions but are chemically unchanged at the end of the reaction.

It is quite difficult to show that a substance is a catalyst in the laboratory. You have to show that:

1. It does speed up the chemical reaction.

2. It is the same chemical at the end of the reaction.

3. It has the same mass at the end of the reaction as at the beginning.

It can be demonstrated by using the decomposition of aqueous hydrogen peroxide. Hydrogen peroxide decomposes to form water and oxygen. The usual catalyst is manganese(IV) oxide.

hydrogen peroxide → water + oxygen
$$2H_2O_2(aq) \rightarrow 2H_2O(l) + O_2(g)$$

Experiment 7.7 The catalytic decomposition of hydrogen peroxide

Weigh out some manganese(IV) oxide (about 1·0 g of the granular form) in a small test tube. Pour about 50 cm^3 of water into a conical flask and then add 10 cm^3 of '10 volume' hydrogen peroxide solution. Carefully place the small test tube inside the flask and connect the gas syringe as shown overleaf.

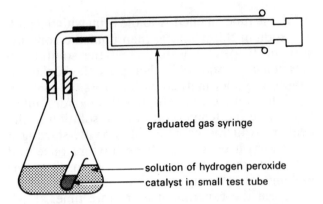

graduated gas syringe

solution of hydrogen peroxide

catalyst in small test tube

Fig 7.11

Tilt the flask so that the test tube falls over. Start a clock and note the time when the syringe reads 10, 20, 30, 40 cm³, etc. You may need to keep swirling the flask to make sure that the manganese(IV) oxide mixes with the aqueous hydrogen peroxide.

Plot a graph of total volume of oxygen (on the vertical axis) against time.

Once again compare your graph with that shown.

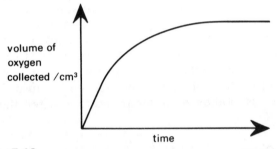

volume of oxygen collected /cm³

time

Fig 7.12

The manganese(IV) oxide increases the speed of reaction because aqueous hydrogen peroxide only decomposes at a very slow rate at room temperature.

To find out if you still have the same amount of the oxide, you must filter it off, wash it, dry it, and then weigh it.

A catalyst is not used up during a reaction, so the mass at the beginning and at the end of the reaction should be the same.

If you have enough time, repeat the experiment with the following changes.

1. Crushing the manganese(IV) oxide to a fine powder before use.
2. Using twice as much catalyst.
3. Warming the mixture of water and hydrogen peroxide to 40°C before adding the catalyst.

If you do not have enough time, sketch the curves you would expect to obtain, using the same axes as before.

Enzymes

These are substances found in nature that are very efficient catalysts. They are compounds with very complicated structures.

Experiment 7.8 The enzymic hydrolysis of starch

To a test tube containing a few cm³ of a freshly prepared 1% starch solution, add a few drops of a solution of iodine in aqueous potassium iodide. Keep this tube to compare with the enzyme reaction.

Mix about 1 cm³ of saliva with a 10 cm³ sample of the starch solution.

Take a small sample from this mixture every five minutes and test it with a few drops of the iodine solution. The hydrolysis is complete when a blue colouration is no longer obtained when the iodine solution is added.

Heating the mixture strongly does not increase the speed of the reaction. This is because most enzymes start to break down at temperatures above 40°C.

The acid hydrolysis of starch is described in Chapter 18.

5. Type of bond

If aqueous silver nitrate and aqueous sodium chloride are mixed, a white precipitate of silver chloride is formed immediately. The speed of reaction is very rapid because the product is formed by oppositely charged ions joining together.

$$Ag^+(aq) + Cl^-(aq) \rightarrow AgCl(s)$$

In the same way as opposite poles of a magnet attract one another, so do oppositely charged particles and, therefore, the reaction is fast. Two particles of the same charge are unlikely to react together rapidly, if at all.

If ethanol is mixed with ethanoic acid there is no apparent reaction. The addition of a catalyst and heating eventually produces a sweet-smelling product. This reaction is much slower than the one quoted above because ethanol and ethanoic acid are covalent compounds and there is no ionic attraction between the reacting molecules.

As a general rule, ionic substances react together more rapidly than covalent substances.

7.3 Reversible reactions

In some chemical reactions the products of the reaction react together to reform the starting substances. These reactions are known as *reversible reactions*.

The reaction between nitrogen and hydrogen to form ammonia is an example of a reversible reaction.

nitrogen + hydrogen $\rightleftharpoons$ ammonia
$$N_2 + 3H_2 \rightleftharpoons 2NH_3$$

Because the reaction is reversible it is impossible

to change all the nitrogen and hydrogen into ammonia.

The $\rightleftharpoons$ sign in the equation tells us that the reaction is reversible. The equations for all reversible reactions have the $\rightleftharpoons$ sign instead of the arrow ($\rightarrow$).

Questions

1. An exothermic reaction is a reaction that:

A produces light
B produces a gas as one of the products
C produces heat energy
D has a solid as one of the products
E has no temperature change.

2. During photosynthesis which one of the following is given out by green plants?

A carbon dioxide
B chlorophyll
C energy
D oxygen
E water vapour.

3. Which one of the following can be deduced from the equation below?
$$CaCO_3(s) + 2HCl(aq) \rightarrow$$
$$CaCl_2(aq) + CO_2(g) + H_2O(l) \text{ exothermic}$$

A Carbon dioxide is soluble in water.
B Calcium chloride is insoluble in water.
C Calcium has been oxidised.
D The total energy of the reactants is greater than the total energy of the products.
E Chlorine has been reduced.

4 Magnesium was reacted with excess hydrochloric acid and the hydrogen produced was collected in a syringe. Which one of the following graphs represents the total volume of hydrogen as measured at various intervals of time?

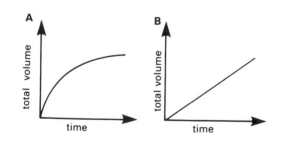

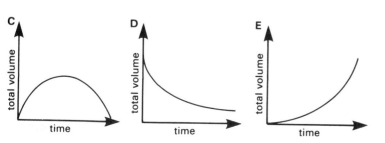

5. Excess calcium carbonate was placed in a flask on a top-pan balance and dilute nitric acid was added. The total mass of the flask and its contents was plotted against time.

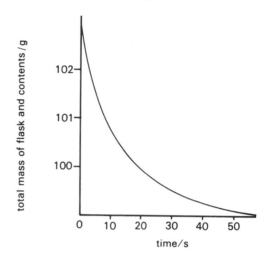

At which one of the following times was the reaction fastest?

A 0s   B 10s   C 20s   D 30s   E 40s

6. Calcium carbonate reacts with dilute hydrochloric acid to form carbon dioxide gas.
In *experiment 1* a lump of calcium carbonate was allowed to react with $50\,cm^3$ of dilute hydrochloric acid. The volume of carbon dioxide formed at various times was noted.

Experiment 2 was a repeat of experiment 1. The *only* difference was that powdered calcium carbonate was used in experiment 2.

The results of experiments 1 and 2 were plotted as the graph shows.

a Which of the curves (A or B) shows the results of experiment 1? Give a reason for your answer.

b Assuming that there is calcium carbonate left at the end of both experiments explain:
(i) why the speed of the reaction gets slower and slower?
(ii) how you can tell from the graph that the same amount of hydrochloric acid was used in experiment 1 and in experiment 2?

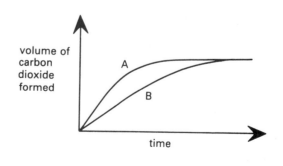

7.

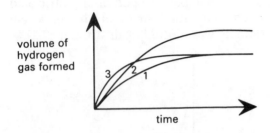

volume of hydrogen gas formed

3   2   1

time

Three separate experiments were performed in which magnesium ribbon was allowed to react with *excess* hydrochloric acid. The amounts of hydrogen formed at different times were noted. The results of these experiments are shown in the graph.

a In which experiment was the most concentrated hydrochloric acid used? Give a reason for your answer.

b In which experiment was the greatest amount of magnesium ribbon used? Give a reason for your answer.

8. Manganese(IV) oxide acts as a catalyst for the decomposition of hydrogen peroxide.

a Write the equation for this decomposition.

b State briefly the general properties associated with catalysts.

c Describe the experiment(s) you would carry out to confirm that manganese(IV) oxide is acting as a catalyst in this decomposition.

d Give the equations for *three* industrial reactions in which catalysts are used. Name the catalyst required for each of your examples.

e Name the catalyst required for the process known as photosynthesis.

[C]

9. Water is formed and energy is released when hydrogen combines with oxygen.

a Write the equation, including the state symbols, for this reaction.

b In this reaction the covalent bonds in the molecules of hydrogen and oxygen are broken.
  (i) Is the bond breaking process exothermic or endothermic?
  (ii) What bonds are formed in the reaction?

c State three forms in which energy may be released.

[C]

10. It is claimed that a finely powdered solid reacts more rapidly than large lumps of solid, when treated with a suitable chemical compound.

a Describe, with the aid of a diagram, how you would carry out experiments to confirm the accuracy of this claim, when considering the reaction of excess dilute hydrochloric acid with: (i) powdered calcium carbonate (ii) lumps of calcium carbonate.

b Sketch the graphs you might expect to obtain if the volumes of carbon dioxide evolved were plotted against the time elapsed from the beginning of the experiment. The two graphs should be on the same axes.

c Comment, with reasons, on the correctness or otherwise of the claim in the first sentence in the question. What other factors might increase the speed of the reaction between calcium carbonate and hydrochloric acid?

[L]

11. A crystal of calcite (pure calcium carbonate) weighing 7·50 g was placed in a flask with 50 cm^3 of dilute hydrochloric acid. The flask was kept at constant temperature and the carbon dioxide evolved was collected in a graduated vessel. The volume of carbon dioxide was recorded at 20-minute intervals. Some of the calcite remained undissolved at the end of the experiment. The results of the experiment are given in the following table:

Time from start of reaction (min)	Volume of carbon dioxide formed at s.t.p. (cm^3)
20	655
40	910
60	1065
80	1100
100	1120
120	1120

a Write the equation for the reaction between calcium carbonate and hydrochloric acid.

b 655 cm^3 of gas were evolved during the first 20-minute interval. Write down the volumes of gas evolved:
  (i) during the second 20-minute interval (20 to 40 minutes)
  (ii) during the third 20-minute interval (40 to 60 minutes)
  (iii) during the fourth 20-minute interval (60 to 80 minutes).
  Explain the variation in the volumes of gas formed during these intervals.

c Why is there no increase in volume of gas after 100 minutes?

d What is the mass of 1120 cm^3 of carbon dioxide at s.t.p.?

e What mass of calcite will have reacted with the acid after 120 minutes?

f Calculate the original concentration of the acid used in grams of hydrogen chloride per dm^3.
  (H = 1; C = 12; O = 16; Cl = 35·5. 1 mole of gas occupies 22·4 dm^3 at s.t.p.)

[C]

8 The mole

8.1 Recipe work

Chemists use recipes to prepare new substances. One of the problems facing a chemist when making up a recipe is: "How much of each ingredient (*reagent*) do I need to use?" Suppose you want to prepare 100 g of copper(II) sulphate crystals starting from copper(II) oxide. What is the smallest amount of copper(II) oxide that you could use? Is it 12·5 g, 32 g, 67 g, 100 g or 160 g?

Guessing the amount of a substance needed for a chemical reaction can be very difficult. Chemists need to be able to work out how much of each substance they need. In this chapter you are going to find out how to do this.

8.2 Relative formula mass

In Chapter 3 you saw that chemists could work out how heavy atoms of each element are by comparing them with 1 atom of carbon ($^{12}_{6}C$). This gave us the relative atomic mass scale. On this scale, carbon (the standard) is given a relative atomic mass of 12. The idea of relative atomic mass has been extended to compounds. You will remember that compounds can exist as molecules or groups of ions.

Chemists have compared the mass of one formula unit of a compound with the mass of one atom of carbon. In this way each substance can be given a relative formula mass (M_r).

Although substances like magnesium oxide do not exist as molecules, we can still use M_r to represent relative formula mass.

For MgO, $M_r = (24 + 16) = 40$.

The relative formula mass of any compound can be worked out by adding together the relative atomic masses (A_r) of each atom in the formula unit.

Example 1. Sodium chloride

Formula NaCl

A_r Na = 23, Cl = 35·5.
$M_r = 23 + 35·5 = 58·5$.

Example 2. Water
Formula H_2O. There are 2 hydrogen atoms and 1 oxygen atom.
A_r H = 1, O = 16.
$M_r = (2 \times 1) + 16 = 18$.

Example 3. Sulphuric acid
Formula H_2SO_4. There are 2 hydrogen atoms, 1 sulphur atom and 4 oxygen atoms.
A_r H = 1, S = 32, O = 16.
$M_r = (2 \times 1) + 32 + (4 \times 16)$
    $= 2 \quad + 32 + 64 = 98$.

Example 4. Ammonium sulphate
Formula $(NH_4)_2SO_4$. Allowing for the brackets there are 2 nitrogen atoms, $2 \times 4 = 8$ hydrogen atoms, 1 sulphur atom and 4 oxygen atoms.
A_r N = 14, H = 1, S = 32, O = 16.
$M_r = (2 \times 14) + (8 \times 1) + 32 + (4 \times 16)$
    $= 28 \quad + 8 \quad + 32 + 64 = 132$.

For COVALENT SUBSTANCES the formula unit is a *molecule*. For these substances the RELATIVE FORMULA MASS is known as the *relative molecular mass* (M_r).

8.3 The mole

Chemists, like cooks, weigh out their ingredients. Chemists cannot count out the number of atoms they want to use in a reaction. For this reason chemists needed to know how heavy each atom was in grams. They have found out that 12 g of carbon ($^{12}_{6}C$) contains six hundred thousand million million million atoms (6×10^{23}). This enormous number is known as Avogadro's constant. To make life simple, we will call this number L.

Consider the four substances shown in Table 1.

Name of substance	Type of formula unit	M_r
carbon	atom	12
magnesium	atom	24
water	molecule	18
sodium chloride	ion pair	58·5

Table 1

L atoms of carbon have a mass of 12 g.
L atoms of magnesium have a mass of 24 g, because each atom of magnesium is twice as heavy as a carbon atom.
L molecules of water have a mass of 18 g, because each water molecule is $1\frac{1}{2}$ times as heavy as a carbon atom.

Once Avogadro's constant had been calculated, chemists were able to work out the number of particles present in any amount of a substance.

Chemists then defined the mole:
A mole is the amount of a substance that contains L formula units of that substance.

So 1 mole of carbon atoms has a mass of 12 g
       1 mole of magnesium atoms has a mass of 24 g
       1 mole of water molecules has a mass of 18 g.

1 mole of any substance has a mass equal to its relative formula mass in grams.

Example: Calcium carbonate.
Formula: $CaCO_3$
There is 1 calcium atom, 1 carbon atom and 3 oxygen atoms

$M_r = 40 + 12 + (3 \times 16)$
$= 40 + 12 + 48 = 100$

So, 1 mole of calcium carbonate has a mass of 100 g.

It is very important to state the particles you are referring to when talking about moles of elements.

For example, the statement 'a mole of oxygen' could mean a mole of oxygen atoms or a mole of oxygen molecules.

You must show, in these cases, what particles you are considering.

Now look at some simple calculations involving Avogadro's constant.

1. If 12 g of carbon contain L atoms, how many atoms are there in 6 g of magnesium?

12 g of carbon is the mass of 1 mole of carbon atoms

6 g of magnesium $= \dfrac{6}{24} = \dfrac{1}{4}$ moles of magnesium atoms

1 mole of magnesium contains L atoms

Therefore $\dfrac{1}{4}$ moles of magnesium contains $\dfrac{L}{4}$ atoms.

2. If the Avogadro constant is 6×10^{23}/mol, how many molecules are there in 22 g of carbon dioxide (CO_2)?

For CO_2, $M_r = 12 + (2 \times 16) = 44$

Therefore 22 g is $\dfrac{22}{44} = \dfrac{1}{2}$ mole of carbon dioxide

Therefore the number of molecules $= \dfrac{6 \times 10^{23}}{2}$
$= 3 \times 10^{23}$

Let us now see how chemists are able to use the idea of the mole.

8.4 Equation calculations

Consider burning magnesium in oxygen to form magnesium oxide. The balanced equation for the reaction tells us that 2 atoms of magnesium react with 1 molecule of oxygen to form 2 formula units of magnesium oxide.

magnesium +	oxygen →	magnesium oxide
2 Mg +	O_2 →	2 MgO
2 atoms	1 molecule	2 formula units

Multiply by L

2L atoms	L molecules	2L formula units

This is the same as

2 moles	1 mole	2 moles
of atoms	of molecules	of formula units

This means

2×24 g	2×16 g	$(2 \times 24) + (2 \times 16)$ g
48 g	32 g	80 g

ie 48 g of magnesium react with 32 g of oxygen to form 80 g of magnesium oxide.

The balanced equation tells us how many particles of each substance react. It also tells us how many moles of each substance react. We therefore know what mass of each substance reacts. This means chemists can calculate the amount of each substance they need for a recipe. There is no need to guess.

We can now think of a chemical equation as showing the number of moles of each substance reacting.

iron +	sulphur →	iron(II) sulphide
Fe +	S →	FeS
1 mole	1 mole	1 mole
of atoms	of atoms	of formula units

copper(II) + oxide	sulphuric → acid	copper(II) + sulphate	water
CuO +	H_2SO_4 →	$CuSO_4$ +	H_2O
1 mole	1 mole	1 mole	1 mole

aluminium +	oxygen →	aluminium oxide
4Al +	$3O_2$ →	$2Al_2O_3$
4 moles	3 moles	2 moles

Now consider some simple calculations involving equations:

1. How much calcium oxide can be obtained from 100 g of calcium carbonate?

The balanced equation for the reaction is:

$CaCO_3$ →	CaO +	CO_2
1 mole	1 mole	1 mole

1 mole of calcium carbonate → 1 mole of calcium oxide

For $CaCO_3$, $M_r = 100$

Therefore, 100 g is 1 mole of $CaCO_3$

So, 1 mole of CaO will be obtained

For CaO, $M_r = 56$

Therefore 56 g of CaO will be obtained.

2. How much magnesium would you have to burn to make 4 g of magnesium oxide?

The balanced equation for the reaction is:

2Mg +	O_2 ⟶	2MgO
2 moles	1 mole	2 moles

For MgO, $M_r = 24 + 16 = 40$

4 g of magnesium oxide is $\dfrac{4}{40} = 0.1$ mole

Since 2 moles of magnesium oxide are obtained from 2 moles of magnesium, 1 mole of magnesium oxide will be obtained from 1 mole of magnesium

Hence 0.1 moles of magnesium oxide will be obtained from 0.1 moles of magnesium

0.1 moles of magnesium $= 24 \times 0.1$
$= 2.4$ g

3. What mass of oxygen is needed to completely burn 32 g methane?

The balanced equation for the reaction is:

$$CH_4 + 2O_2 \rightarrow CO_2 + 2H_2O$$
1 mole   2 moles   1 mole   2 moles

Mass of 1 mole of methane $= 12 + (4 \times 1) = 16\,g$

$$32\,g \text{ of methane} = \frac{32}{16} = 2 \text{ moles}$$

Since 1 mole of methane reacts with 2 moles of oxygen

then 2 moles of methane will react with 4 moles of oxygen.

Mass of 1 mole of oxygen molecules $= 2 \times 16 = 32\,g$
Hence mass of oxygen required $= 4 \times 32\,g$
$$= 128\,g$$

Calculations involving volumes of gases

It is easier to find out how much gas is present by measuring its volume rather than its mass. It is useful if chemists can calculate the volume of gas they expect to be formed in a chemical reaction.

Avogadro's Law makes this possible.

Avogadro's Law states that equal volumes of all gases under the same conditions of temperature and pressure contain the same number of molecules.

It follows from this that the volume occupied by 1 mole of gas under the same conditions of temperature and pressure is the same for all gases.

1 mole of a gas occupies $22400\,cm^3$ at $0\,°C$ and 1 atmosphere pressure. (These conditions are called standard temperature and pressure and they are usually abbreviated to s.t.p.)

1 mole of a gas will occupy $24000\,cm^3$ at $20\,°C$ and 1 atmosphere pressure. (These conditions are called room temperature and pressure (r.t.p.).)

The volume occupied by 1 mole of a gas is called its *molar volume*.

Consider the following calculation:

What is the maximum volume of carbon dioxide, measured at room temperature and pressure, that can be obtained from $4.2\,g$ of magnesium carbonate when it reacts with an excess of dilute sulphuric acid?

The balanced equation for this reaction is:

$$MgCO_3 + H_2SO_4 \rightarrow MgSO_4 + CO_2 + H_2O$$
1 mole   1 mole   1 mole   1 mole   1 mole

1 mole of magnesium carbonate $\rightarrow$ 1 mole of carbon dioxide

For $MgCO_3$, $M_r = 24 + 12 + (3 \times 16) = 84$

1 mole of carbon dioxide at room temperature and pressure occupies $24000\,cm^3$

Therefore
$$84\,g \text{ of } MgCO_3 \rightarrow 24000\,cm^3 \text{ of } CO_2$$

$$4.2\,g \text{ of } MgCO_3 \text{ is } \frac{4.2}{84} = 0.05 \text{ moles}$$

Therefore 0·05 moles of CO_2 are given off
0.05 moles of $CO_2 = 0.05 \times 24000\,cm^3$
$$= 1200\,cm^3$$
So $1200\,cm^3$ of carbon dioxide are produced.

8.5 Calculating formulae

We can use the idea of moles to calculate the formula of a substance from experimental results. The formula calculated is the simplest possible formula for that compound. It is known as the empirical formula of the substance.

Example 1: A compound is formed by 32 g of sulphur reacting with 32 g of oxygen. What is the formula of the compound formed? $(S = 32, O = 16)$

Mass of 1 mole of sulphur atoms $= 32\,g$
Mass of 1 mole of oxygen atoms $= 16\,g$

The experimental results tell us that:

1 mole of sulphur atoms react with 2 moles of oxygen atoms
L atoms of sulphur react with $2L$ atoms of oxygen.
1 atom of sulphur reacts with 2 atoms of oxygen
So, the empirical formula of the compound must be SO_2.

Example 2: 3 g of carbon react with 8 g of oxygen to form a compound. What is the formula of this compound? $(C = 12, O = 16.)$

$$3\,g \text{ of carbon} = \frac{3}{12} \text{ moles} = \tfrac{1}{4} \text{ mole of carbon atoms}$$

$$8\,g \text{ of oxygen} = \frac{8}{16} = \tfrac{1}{2} \text{ mole of oxygen atoms}$$

$\tfrac{1}{4}$ mole of carbon atoms react with $\tfrac{1}{2}$ mole of oxygen atoms

$4 \times \tfrac{1}{4} = 1$ mole of carbon atoms react with $4 \times \tfrac{1}{2} = 2$ moles of oxygen atoms.

L atoms of carbon react with $2L$ atoms of oxygen
1 atom of carbon reacts with 2 atoms of oxygen

So, the empirical formula of the compound formed must be CO_2.

Experiment 8.1 To find the formula of magnesium oxide

Weigh a crucible and lid. Take a piece of magnesium ribbon 15cm long, and scrape the surface to remove the oxide layer. Coil the magnesium and place in the crucible. Place the lid on the crucible and weigh again. Heat the crucible gently on a pipe-clay triangle. When the magnesium catches fire, heat it more strongly. Slightly lift the crucible lid using tongs to allow a little air into the crucible. Quickly replace the lid making sure that no magnesium oxide powder is lost. Repeat this process until the magnesium ceases to flare up. Remove the crucible lid and heat the crucible strongly to make sure that the magnesium has reacted completely. Allow the crucible to cool, replace the lid and weigh. Repeat the heating process, cooling and weighing until a constant mass is obtained. Record your results as shown:

Mass of crucible + lid = g
Mass of crucible + lid + magnesium = g
Mass of crucible + lid + magnesium oxide = g

From your results, calculate the simplest formula of magnesium oxide.

8.6 % yield and % purity

Chemists often want to know the purity of a substance, or the expected yield of a reaction. The following calculations show how these values can be found.

1. When 6 g of carbon containing impurities was burnt in excess oxygen, 16·5 g of carbon dioxide was obtained. Calculate the % purity of the carbon.

The balanced equation for the reaction is:

$$\underset{\text{1 mole}}{C} + \underset{\text{1 mole}}{O_2} \rightarrow \underset{\text{1 mole}}{CO_2}$$

For CO_2, $M_r = 44$

Number of moles of carbon dioxide formed

$$= \frac{16\cdot5}{44} = 0\cdot375 \text{ moles}$$

Since 1 mole of carbon dioxide is obtained from 1 mole of carbon, the number of moles of carbon in the original sample must be 0·375 moles.

Mass of 1 mole of carbon = 12 g

Hence the mass of carbon in the original sample $= 0\cdot375 \times 12 = 4\cdot5$ g

$$\% \text{ purity} = \frac{4\cdot5}{6} \times 100 = 75\% \text{ pure}$$

2. The reaction between ethanol and ethanoic acid gives ethyl ethanoate and water.

$$C_2H_5OH + CH_3CO_2H \rightleftharpoons CH_3CO_2C_2H_5 + H_2O$$

What is the percentage yield of ethyl ethanoate if 2·2 g of ethyl ethanoate was obtained by reacting 4·6 g of ethanol with excess ethanoic acid.

Mass of 1 mole of ethanol = 46 g

Number of moles of ethanol used $= \frac{4\cdot6}{46} = 0\cdot1$

Expected number of moles of ethyl ethanoate to be formed $= 0\cdot1$

Mass of 1 mole of ethyl ethanoate = 88 g

Therefore, expected mass of ethyl ethanoate to be formed $= 0\cdot1 \times 88 = 8\cdot8$ g

$$\% \text{ yield} = \frac{2\cdot2}{8\cdot8} \times 100$$

$$= 25\%$$

Questions

1. The compound magnesium sulphide is made by heating magnesium with sulphur. The two elements combine together in the ratio of 3 parts of magnesium to 4 parts of sulphur. If 8 g of magnesium was heated with 8 g of sulphur, the composition of the product would be:

A 7 g of magnesium sulphide
B 8 g of magnesium sulphide
C 16 g of magnesium sulphide
D 7 g of magnesium sulphide and 8 g of sulphur
E 14 g of magnesium sulphide and 2 g of magnesium.

2. When manganese(II) ethanedioate is heated in a crucible, an oxide of manganese is formed containing 63·2% manganese and 36·8% oxygen. What is the formula of the oxide obtained?

A MnO
B Mn_2O_3
C MnO_2
D MnO_3
E Mn_2O_7

3. What is the maximum mass of aluminium that can be obtained from 34 g of pure aluminium oxide (Al_2O_3)?

A 9 g
B 18 g
C 27 g
D 36 g
E 54 g

4. 20 cm^3 of an oxide of nitrogen was completely decomposed into 20 cm^3 of a mixture of nitrogen and oxygen. When oxygen was removed from the mixture, 10 cm^3 of nitrogen remained. All volumes were measured at the same temperature and pressure. Which one of the following is the formula of this oxide?

A N_2O
B NO
C N_2O_3
D NO_2
E N_2O_5

5. 41 g of hydrated magnesium sulphate crystals $(M_r = 246)$ were obtained when 8 g of magnesium $(A_r = 24)$ was reacted with excess dilute sulphuric acid. What is the percentage yield of the hydrated magnesium sulphate?

A 20%
B 30%
C 40%
D 50%
E 60%

6. Calculate the relative formula mass of the following:

a CO_2 b $CaCO_3$ c FeS
d $CuSO_4$ e C_6H_{12} f H_2O

g $Ca(OH)_2$ **h** O_2 **i** $(NH_4)_2SO_4$
j Na_2O

7. Calculate the mass of 1 mole of the following:

a Cl_2 **b** SO_2 **c** Na_2CO_3
d CuO **e** $Zn(NO_3)_2$ **f** $CaCl_2$
g $MgCO_3$ **h** MnO_2 **i** C_2H_6
j $C_2H_4O_2$

8. $2Pb(NO_3)_2(s) \xrightarrow{\text{heat}} 2PbO(s)$
$$+ 4NO_2(g) + O_2(g)$$

Use the above equation to answer the following questions:

a How many moles of lead(II) nitrate have to be heated to make 4 moles of nitrogen dioxide?
b If 4 moles of lead(II) nitrate are heated how many moles of oxygen gas will be formed?
c If 10 moles of lead(II) oxide are formed how many moles of oxygen gas will be formed at the same time?
d Calculate the mass of: (i) 1 mole of oxygen gas, (ii) 4 moles of nitrogen dioxide gas
e If $4 \, dm^3$ of nitrogen dioxide is formed in this reaction what volume of oxygen is formed?

9. In each of the following calculate the formula of the compound formed when:

a 24 g of carbon react with 32 g of oxygen
b 46 g of sodium react with 16 g of oxygen
c 12 g of magnesium react with 80 g of bromine
d 6 g of carbon react with 2 g of hydrogen
e 10 g of hydrogen react with 80 g of oxygen
f 24 g of sulphur react with 36 g of oxygen.

10. The equation for the thermal decomposition of sodium hydrogencarbonate is:

$2NaHCO_3(s) \rightarrow$
$$Na_2CO_3(s) + H_2O(g) + CO_2(g)$$

a Explain what is meant by *thermal decomposition*.
b If 21 g of sodium hydrogencarbonate were decomposed, calculate: (i) the mass of the residue formed (ii) the volume of carbon dioxide evolved, measured at s.t.p.
c Ethyl ethanoate $(CH_3CO_2C_2H_5)$ can be prepared from ethanol and ethanoic acid.
(i) Write the equation for the reaction.
(ii) Calculate the theoretical yield of ethyl ethanoate which can be obtained from 23 g of ethanol.
(iii) Experimentally it was found that 33 g of ethyl ethanoate were obtained from 23 g of ethanol. Calculate the percentage yield.

[C]

11. 4 g of copper completely reacted with 1 g of sulphur to form a black solid X. When 5 g of X were heated in an excess of oxygen, sulphur dioxide was given off and 5 g of a black solid Z remained.

a Calculate the simplest formula of X.
b Identify Z, giving reasons for your answer.
c Write a balanced equation for the reaction between X and oxygen.
d How does your equation in **c** account for the fact that the mass of X and the mass of Z are the same?
(Relative atomic masses: O, 16; S, 32; Cu, 64.)

[C]

12. Strontium (Sr) is a reactive metal with a valency of 2.

Phosphorus (P) is a reactive non-metal with a relative atomic mass of 31.

Both strontium and phosphorus burn in oxygen to give oxides and both of these oxides react with water to give soluble compounds.

The oxide of phosphorus contains 56·3% by mass of oxygen.

Strontium phosphide—formula Sr_3P_2—is a solid with a high melting point. It reacts with water to give a gas of formula PH_3 and a strontium compound X as the only products.

a Write the formula for strontium oxide and *calculate* the formula for the oxide of phosphorus from the percentage given above.
b Write equations for the reactions you would expect between water and *each* of the above oxides. Give a simple test by which you could distinguish between the solutions formed.
c Write the formula you would expect for compound X and the equation for the reaction of compound Sr_3P_2 with water.
d Calculate the volume at s.t.p. of the gas PH_3 (which is insoluble in water) formed when one mole of Sr_3P_2 reacts with an excess of water and the volume of $1·0 \, mol/dm^3$ hydrochloric acid required to neutralise the resulting solution.
e Name the type of bonding you would expect in compound Sr_3P_2. Give a reason for your answer.
f What valencies does phosphorus show in the above reactions?

[C]

13. This question is about an experiment to find the equation for the reaction between a colourless solution containing carbonate ions, $CO_3^{2-}(aq)$, and a green solution containing nickel ions, $Ni^{2+}(aq)$, to form a precipitate of green nickel carbonate. $5 \, cm^3$ of $1·0 \, mol/dm^3$ sodium carbonate solution, $Na_2CO_3(aq)$, was measured into a test tube, $1 \, cm^3$ of $1·0 \, mol/dm^3$ nickel chloride solution, $NiCl_2(aq)$, was measured into the same test tube, the tube

was shaken to mix the reagents, and the precipitate formed was allowed to settle evenly. The height of the precipitate was measured.

A second 1 cm³ portion of the nickel chloride solution was measured into the same tube and, after shaking to mix and allowing the precipitate to settle evenly, the height of the precipitate was measured.

This was repeated for successive additions of 1 cm³ of the 1·0 mol/dm³ nickel chloride solution. A graph of the results is given below.

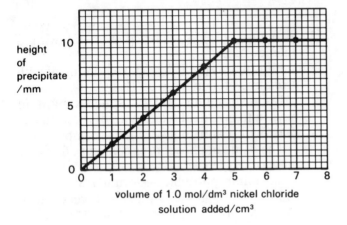

height of precipitate /mm

volume of 1.0 mol/dm³ nickel chloride solution added/cm³

a   Why did the height of the precipitate not rise after 5 cm³ of nickel chloride solution had been added?

b   What would be another observable difference (other than the height of the precipitate) in the contents of the test tube before and after the fifth 1 cm³ portion had been added?

c   In what ratio of moles of ions did the reagents react?

$$\frac{\text{Moles of } CO_3{}^{2-}(aq)}{\text{Moles of } Ni^{2+}(aq)} = \underline{\hspace{3cm}}$$

d   Using the information given above, write an ionic equation for the reaction.

e   If 0·5 mol/dm³ sodium carbonate solution were used, what volume of it would be required to produce the same mass of nickel carbonate as formed in the experiment?

[L]

9 Acids, bases and salts

9.1 Recognising acid and alkalis

When we first think of acids we think of liquids with a sour taste that burn the skin. When we first think of alkalis we think of liquids that neutralise acids and make our skin feel soapy. It would be silly to decide whether a substance was an acid or an alkali by tasting it, or pouring it on your hand. Chemists have had to find a different method.

Chemists find out if a substance is an acid or an alkali by adding an indicator to a solution of the substance.

An indicator is a substance which usually has one colour in very acidic solutions and another colour in very alkaline solutions.

The following table shows some common indicators:

Name of indicator	Colour in acidic solution	Colour in alkaline solution
Litmus	Red	Blue
Phenolphthalein	Colourless	Pink
Methyl orange	Red	Yellow
Screened methyl orange	Red	Green

Table 1

Litmus is the most common indicator.

These indicators tell us if a solution is acidic or alkaline, but they do not tell us how acidic or how alkaline the solution is.

pH scale

This is a number scale used to describe how acidic or how alkaline a solution is.

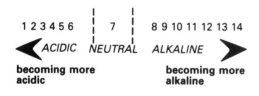

The pH of solutions is usually measured by using Universal indicator. (This is a mixture of indicators.) It has different colours in solutions of different pH.

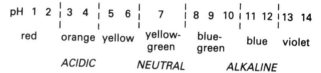

9.2 Acids

Experiment 9.1 Properties of acids

You will need the following acids: dilute sulphuric acid, dilute hydrochloric acid, dilute nitric acid, a solution of citric acid in water and solid citric acid.

In each of the following three tests use about 2 cm depth of acid solution in a test tube.

1. Add a small piece of Universal indicator paper.
2. Add about 2 cm of magnesium ribbon. Test the gas given off for hydrogen.
3. Add a small spatula measure of sodium carbonate. Test the gas given off for carbon dioxide.
4. Using the apparatus shown in Fig 10.2, test each solution to see if it conducts an electric current.

Repeat the tests **1** to **3** using solid citric acid.

Table 2 gives information about some common acids.

The table tells us that:
1. Acids contain hydrogen.
2. Acids can be solids, liquids or gases.

Name of acid	Molecular formula	Where can it be found?	State at room temperature (when pure)
Hydrochloric	HCl	In your stomach	Gas
Sulphuric	H_2SO_4	In car batteries	Liquid
Nitric	HNO_3	Formed in atmosphere during thunderstorms	Liquid
Ethanoic	CH_3CO_2H	In vinegar	Liquid
Citric	$H_8C_6O_7$	In lemon juice	Solid

Table 2

Properties of acids

Note Acids only show these properties if water is present.

PROPERTY	EXCEPTION
1. Have a sour taste.	
2. Turn litmus red.	
3. Turn Universal indicator red, orange or yellow.	
4. Produce carbon dioxide from carbonates.	Sometimes the reaction stops after a short while.
5. Produce hydrogen gas with most metals.	Nitric acid does not Unreactive metals like Cu, Ag do not react.
6. Dissolve in water to form solutions which conduct electricity.	

Table 3

The basicity of acids

Sodium chloride (NaCl) is made from HCl by replacing the hydrogen atom by a sodium atom. Replacing the hydrogen atom in HNO_3 by K gives KNO_3 (potassium nitrate). Acids such as HCl and HNO_3 that have *one* 'replaceable' hydrogen atom per molecule are called *monobasic* acids.

Zinc sulphate ($ZnSO_4$) is formed when zinc reacts with dilute sulphuric acid—both hydrogen atoms in H_2SO_4 having been replaced. Acids with *two* 'replaceable' hydrogen atoms are called *dibasic* acids.

Although ethanoic acid (CH_3CO_2H) contains four hydrogen atoms, only one of these is replaceable (CH_3CO_2Na can be made), so this acid is monobasic.

9.3 Bases and alkalis

A simple definition of bases is that they are the oxides and hydroxides of metals e.g. CuO, $Ca(OH)_2$.

Alkalis are simply defined as bases that dissolve in water, eg NaOH.

The following table will give you some information about common alkalis.

Name of alkali	Formula	State at room temperature (when pure)
Sodium hydroxide	NaOH	Solid
Potassium hydroxide	KOH	Solid
Calcium hydroxide	$Ca(OH)_2$	Solid
Barium hydroxide	$Ba(OH)_2$	Solid
Ammonia	NH_3	Gas

Table 4

With the exception of ammonia, all alkalis are metal hydroxides. The metals which form alkalis are in Groups I and II of the Periodic Table.

There is a link between the Periodic Table and the acidity or basicity of the oxides of the elements. As we go across the period from sodium to chlorine, the oxides change from basic to acidic. This change occurs because the character of the elements themselves change from metallic to non-metallic as shown in Table 5.

Atomic number	11	12	13	14	15	16	17
Symbol of element	Na	Mg	Al	Si	P	S	Cl
Formula of an oxide	Na_2O	MgO	Al_2O_3	SiO_2	P_2O_5	SO_3	Cl_2O_7
pH of aqueous solution	14	9	7 (insoluble)	7 (insoluble)	4	3	2
Reaction of oxide	react with acids		reacts with acids and alkalis		react with bases		
Type of oxide		BASIC	AMPHOTERIC			ACIDIC	

Table 5

Those metal oxides (and hydroxides) that react with both acids and alkalis are known as *amphoteric oxides* (or *hydroxides*).

The only common amphoteric oxides are: aluminium oxide, lead(II) oxide, zinc oxide.

They react with acids and alkalis to form a salt.

Zinc oxide reacts with sulphuric acid as follows:

zinc oxide + sulphuric acid → zinc sulphate + water

It reacts with sodium hydroxide:

zinc oxide + sodium hydroxide → sodium zincate

Aluminium oxide reacts with sodium hydroxide to form sodium aluminate.

Experiment 9.2 Properties of alkalis

You will need: dilute aqueous sodium hydroxide, dilute aqueous potassium hydroxide, lime-water (aqueous calcium hydroxide), a solution of ammonia in water.

Remember to wear safety glasses at all times.

In each of the following tests use about 2 cm depth of alkali solution in a boiling tube.

1. Add a small piece of Universal Indicator paper.
2. Add a small spatula measure of solid ammonium chloride. Warm the mixture. Test the gas given off for ammonia. (Do not do this experiment with the solution of ammonia in water.)
3. Add a small spatula measure of aluminium powder. Warm the solution gently. Test the gas given off for hydrogen. (**Take care, heat gently and use safety glasses.**)
4. Using the apparatus shown in Fig 10.2, test each solution to see if it conducts an electric current.

Properties of alkalis

Note Alkalis only show these properties if water is present.

PROPERTY	EXCEPTION
1. Have a soapy feel.	
2. Turn litmus blue.	
3. Turn Universal indicator blue-green, blue or violet.	
4. Produce hydrogen gas with aluminium.	Not observed with aqueous ammonia or aqueous calcium hydroxide.
5. Dissolve in water forming conducting solutions.	
6. React with acids forming a salt and water only.	
7. React with ammonium salts forming ammonia.	

Table 6

Experiment 9.3 Investigating the properties of hydrogen chloride dissolved in methylbenzene

Repeat the tests described in Experiment 9.1 with a solution of hydrogen chloride dissolved in methylbenzene.

Compare the properties of a solution of hydrogen chloride in water (hydrochloric acid) with the properties of a solution of hydrogen chloride in methylbenzene.

Dip a thermometer bulb into methylbenzene and then into hydrogen chloride gas. Note any change in temperature.

Dry the thermometer bulb and this time dip it into water before placing it in the hydrogen chloride gas. Again note if there is any change in temperature.

You should have observed that when dissolved in methylbenzene, hydrogen chloride does not act as an acid. In this solvent the hydrogen chloride is present as HCl molecules. When dissolved in water the molecules break up into H^+ and Cl^- ions. Acidity is caused by the presence of H^+ ions.

By studying a large number of acids and alkalis chemists have been able to find out why acids have properties in common and why alkalis have properties in common. See Table 7.

It can be shown that all alkaline solutions contain hydroxide ions (OH^-). Metal oxides that dissolve in water produce hydroxide ions.

Example:
$$Na_2O(s) + H_2O(l) \rightarrow 2Na^+(aq) + 2OH^-(aq)$$

Neutralisation

When an acid is neutralised by an alkali, the hydrogen ions react with hydroxide ions to form water molecules.

$$H^+(aq) + OH^-(aq) \rightarrow H_2O(l)$$

When an acidic solution is neutralised its pH increases until it reaches 7. Any solution of pH = 7 is neutral.

Strong and weak acids

We have said that when an acid dissolves in water, the acid molecules change into ions. *Example:*

$$HCl(g) + (aq) \rightarrow H^+(aq) + Cl^-(aq)$$

In a *strong* acid *most of* the molecules change into ions when dissolved in water.

Example: hydrochloric acid, sulphuric acid

In a *weak* acid only some of the molecules change into ions when the acid is dissolved in water.

Example: ethanoic acid.

$$CH_3CO_2H(aq) \rightleftharpoons CH_3CO_2^-(aq) + H^+(aq)$$

A $0.1\,mol/dm^3$ solution of hydrochloric acid has a pH of 1. A $0.1\,mol/dm^3$ solution of ethanoic acid has a pH of 3.

Acids

OBSERVATION	DEDUCTION
1. Acids do not show acidic properties when pure or when dissolved in solvents other than water.	**1.** Water must play an important part in acidity.
2. Acids do not conduct electricity when pure or dissolved in solvents other than water.	**2.** Acids are covalent substances.
3. Water is a poor conductor of electricity.	**3.** Water is a covalent substance.
4. All acids conduct electricity when dissolved in water.	**4.** Acids must form ions when dissolved in water.
5. All solutions of acids in water produce hydrogen gas at the negative electrode when electricity is passed through the solution.	**5.** All acidic solutions must contain *positive* hydrogen ions.

From this type of reasoning, we can say that acids are substances that produce H^+ ions when added to water.

Alkalis

OBSERVATION	DEDUCTION
1. Pure alkalis do not show alkaline properties, but solutions of alkalis dissolved in water do.	**1.** Water plays an important part in alkaline behaviour.
2. All solutions of alkalis in water conduct electricity.	**2.** Alkaline solutions must contain ions. An ion must cause alkaline behaviour.

Table 7

Strong and weak alkalis

A *strong* alkali is almost *completely* ionised when dissolved in water.

Example: NaOH.

A *weak* alkali *produces relatively few ions* when dissolved in water.

Example: ammonia.

Although aqueous ammonia is a common alkali, ammonia does not fit our definition of a base. *A better definition of bases is that they can accept protons. An acid is a substance that can donate a proton:*

$$HCl(g) + H_2O(l) \rightarrow H_3O^+(aq) + Cl^-(aq)$$

The hydrogen chloride molecule has donated a proton to a water molecule to form $H_3O^+(aq)$ (which is really the same as $H^+(aq)$). The water molecule is acting as a base. The following equation shows a water molecule acting as an acid:

$$NH_3(g) + H_2O(l) \rightleftharpoons NH_4^+(aq) + OH^-(aq)$$

The ammonia molecule acts as a base in accepting a proton from the water molecule. In the reverse reaction the ammonium ion acts as an acid and the hydroxide ion as a base.

Metal oxides act as bases because the oxide ion accepts two protons to form a water molecule:

$$Cu^{2+}O^{2-}(s) + 2H^+(aq) \rightarrow Cu^{2+}(aq) + H_2O(l)$$

9.4 Salts

Salts are ionic substances. They contain the positive ions of a metal (or ammonium ions) and the negative ions formed when an acid dissolves in water (see Table 8).

We can make salts in the laboratory in a number of ways.

METHODS OF MAKING SALTS

A. Methods which use acids

1. By reacting a metal with an acid

Many metals react with acids to produce salts and hydrogen gas:

Example:

METAL + ACID → SALT + HYDROGEN

zinc + sulphuric → zinc + hydrogen
          acid   sulphate

$$Zn(s) + H_2SO_4(aq) \rightarrow ZnSO_4(aq) + H_2(g)$$

Experiment 9.4 To prepare magnesium sulphate crystals

Method

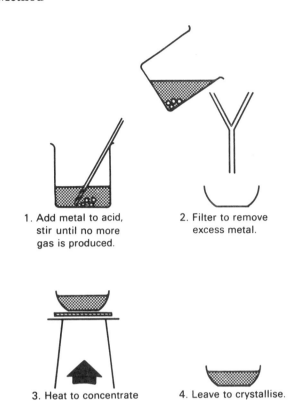

1. Add metal to acid, stir until no more gas is produced.

2. Filter to remove excess metal.

3. Heat to concentrate the salt solution.

4. Leave to crystallise.

Fig 9.1

Put about 25 cm³ of bench sulphuric acid into a small beaker. Add magnesium ribbon in about 1 cm lengths, until no more will react. Stir the solution after adding each piece of magnesium.

Heat the solution to boiling point, and filter the solution whilst hot. Excess magnesium ribbon is left on the filter paper.

Heat the magnesium sulphate solution to evaporate water. Continue boiling the solution until the solution is concentrated enough to crystallise.

Test the solution by removing a few drops on a glass rod, note if crystals form when the solution cools. Cover the hot saturated solution and leave to cool. Slow cooling produces good crystals. Filter them off and wash with a small amount of cold distilled water. Dry the crystals between filter papers.

Name of salt	Formula	Positive ion	Negative ion
magnesium sulphate	$MgSO_4$	Mg^{2+}	SO_4^{2-}
copper(II) chloride	$CuCl_2$	Cu^{2+}	Cl^-
lead(II) chloride	$PbCl_2$	Pb^{2+}	Cl^-
ammonium nitrate	NH_4NO_3	NH_4^+	NO_3^-

Table 8

When not to use this method

If the metal in the salt is very reactive, eg sodium.
If the metal does not react with an acid (ie very unreactive metals such as copper or silver).
If the salt does not dissolve in water.

2. By reacting a metal carbonate with an acid

Method

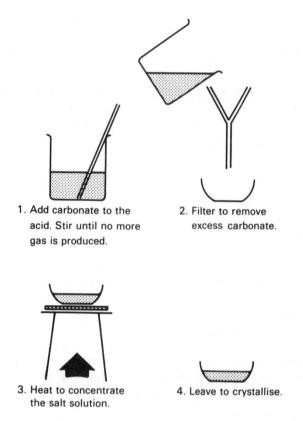

1. Add carbonate to the acid. Stir until no more gas is produced.

2. Filter to remove excess carbonate.

3. Heat to concentrate the salt solution.

4. Leave to crystallise.

Fig 9.2

Experiment 9.5 To prepare lead(II) nitrate crystals

Put about 25 cm³ of bench nitric acid into a small beaker. Add lead(II) carbonate a spatula measure at a time, until no more reacts. Stir the solution after adding each spatula load. Filter off the excess lead(II) carbonate and obtain crystals of lead(II) nitrate as described in Experiment 9.4.

Write the equation for the preparation of lead(II) nitrate by this method.

Metal carbonates react with acids to form a salt, carbon dioxide gas and water:

CARBONATE + ACID → SALT + CARBON DIOXIDE + WATER
copper(II) + sulphuric → copper(II) + carbon dioxide + water
carbonate   acid   sulphate
$CuCO_3(s)$ + $H_2SO_4(aq)$ → $CuSO_4(aq)$ + $CO_2(g)$ + $H_2O(l)$

When not to use this method

If the salt does not dissolve in water.
If the carbonate *does* dissolve in water.

3. By reacting a base with an acid.

A base is a substance that reacts with an acid to form a salt and water only. They are usually metal oxides or metal hydroxides.

a For bases that do not dissolve in water

Example:

ACID + BASE → SALT + WATER
sulphuric + copper(II) → copper(II) + water
acid   oxide   sulphate
$H_2SO_4(aq)$ + $CuO(s)$ → $CuSO_4(aq)$ + $H_2O(l)$

Method

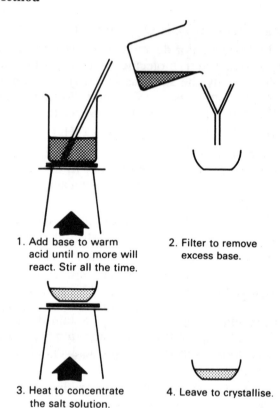

1. Add base to warm acid until no more will react. Stir all the time.

2. Filter to remove excess base.

3. Heat to concentrate the salt solution.

4. Leave to crystallise.

Fig 9.3

Experiment 9.6 To prepare copper(II) sulphate crystals

The method used is the same as in Experiment 9.5, but in this experiment you use bench sulphuric acid and copper(II) oxide.

When not to use this method

If the base is an alkali.
If the salt is insoluble in water.

b *Reaction between an alkali (soluble base) and an acid*

Example:

sodium hydroxide	+	hydrochloric acid	→	sodium chloride	+	water
NaOH(aq)	+	HCl(aq)	→	NaCl(aq)	+	$H_2O(l)$

Method

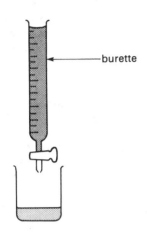

← burette

Fig 9.4

Experiment 9.7 To prepare sodium chloride

Fill a burette with hydrochloric acid up to the zero mark. (Bottom of curve of liquid, called the meniscus, should be level with the zero mark—see diagram.) Use a pipette to place $25\,cm^3$ of sodium hydroxide in the beaker. Add two drops of indicator to the sodium hydroxide.

Run acid from the burette into the alkali until the solution *just* changes colour. Note the volume of hydrochloric acid used.

Repeat the experiment using the same volume of hydrochloric acid and sodium hydroxide but with no indicator. Heat the solution in an evaporating basin until most of the liquid has evaporated. Leave to crystallise.

B. Methods not needing acids

1. *Precipitation method*

Method

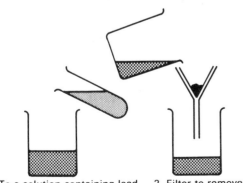

1. To a solution containing lead ions add a solution containing chloride ions until no more precipitate forms.
2. Filter to remove the precipitate.

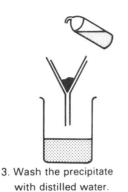

3. Wash the precipitate with distilled water.
4. Dry the precipitate.

Fig 9.5

This is used only to prepare salts which do not dissolve in water. The salt is formed as a precipitate:

Example:

lead(II) nitrate	+	sodium chloride	→	lead(II) chloride	+	sodium nitrate
$Pb(NO_3)_2(aq)$	+	$2NaCl(aq)$	→	$PbCl_2(s)$	+	$2NaNO_3(aq)$

Experiment 9.8 To prepare lead(II) sulphate

Put about $10\,cm^3$ of lead(II) nitrate solution into a small beaker. Add excess bench sulphuric acid. Filter to remove the lead(II) sulphate. Wash the precipitate with small amounts of distilled water. Leave the solid to dry.

Write the equation for the preparation of lead(II) sulphate by this method.

The following table gives information about the solubility of salts and bases.

TYPE OF COMPOUND	SOLUBILITY
CARBONATE	All INSOLUBLE *except* sodium, potassium and ammonium
CHLORIDE	All SOLUBLE *except* silver and lead.
HYDROXIDE (and OXIDE)	All INSOLUBLE *except* sodium, potassium, calcium and ammonium.
NITRATE	All soluble.
SULPHATE	All soluble *except* barium, calcium and lead.

Table 9

2. *Synthesis method*

Synthesis is the building up of a compound from simpler substances, usually its elements.

Example:

metal element	+	non-metal element	→	salt
zinc	+	sulphur	→	zinc sulphide
Zn(s)	+	S(s)	→	ZnS(s)

Experiment 9.9 To prepare iron(II) sulphide

Mix a spatula measure of sulphur with a spatula measure of 'grease free' iron filings. Heat the mixture on flame resistant paper, holding the paper in a pair of tongs.

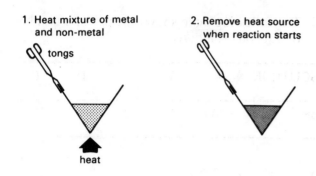

1. Heat mixture of metal and non-metal

2. Remove heat source when reaction starts

tongs

heat

Fig 9.6

Write the equation for the preparation of iron(II) sulphide by this method.

Water of crystallisation

Many substances contain 'hidden water'. The substances are perfectly dry but contain water which is chemically held. This chemically-held water is known as *water of crystallisation*. Copper(II) sulphate crystals and sodium carbonate crystals are examples of substances with water of crystallisation.

Copper(II) sulphate crystals have the formula $CuSO_4.5H_2O$. Each formula unit of copper(II) sulphate has 5 water molecules stuck onto it. Sodium carbonate crystals have the formula $Na_2CO_3.10H_2O$.

Substances that contain this water of crystallisation are known as *hydrated* substances. When they are heated they lose their water. They are then known as *anhydrous* substances.

$$
\begin{array}{ccc}
\text{hydrated} & & \text{anhydrous} \\
\text{copper(II)} & \rightarrow & \text{copper(II)} + \text{water} \\
\text{sulphate} & & \text{sulphate}
\end{array}
$$

$$CuSO_4.5H_2O \rightarrow CuSO_4 + 5H_2O$$

It is possible to collect some of the water hidden in hydrated substances by using the apparatus shown in Fig 9.8.

How would you find out if the water collected is pure water?

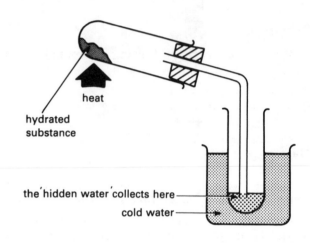

heat

hydrated substance

the 'hidden water' collects here

cold water

Fig 9.8 Collecting 'hidden water'

CHOOSING A METHOD OF PREPARING A SALT

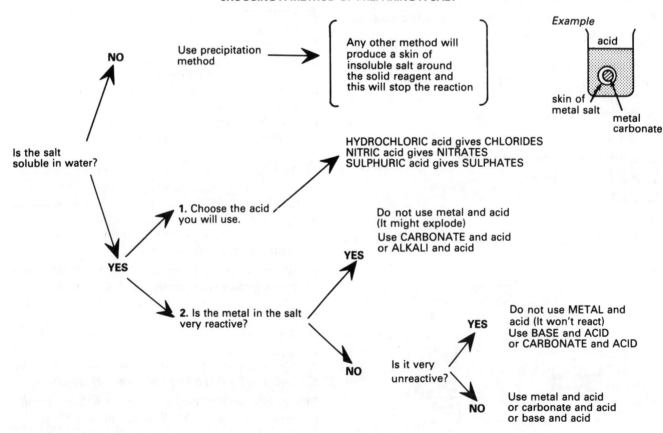

NO

Use precipitation method → Any other method will produce a skin of insoluble salt around the solid reagent and this will stop the reaction

Example

acid

skin of metal salt

metal carbonate

Is the salt soluble in water?

YES

1. Choose the acid you will use.

HYDROCHLORIC acid gives CHLORIDES
NITRIC acid gives NITRATES
SULPHURIC acid gives SULPHATES

Do not use metal and acid (It might explode)
Use CARBONATE and acid or ALKALI and acid

YES

2. Is the metal in the salt very reactive?

NO

Is it very unreactive?

YES

Do not use METAL and acid (It won't react)
Use BASE and ACID or CARBONATE and ACID

NO

Use metal and acid or carbonate and acid or base and acid

Fig 9.7

Some hydrated substances do not require heating to lose their water of crystallisation. Hydrated sodium carbonate is a crystalline solid, but on exposure to air it loses its water of crystallisation and turns into a white powder. This process is known as *efflorescence*.

Other substances behave in the opposite way, they absorb water from the air. Calcium chloride will absorb so much water vapour that a solution is eventually formed. Such a substance is *deliquescent*. Concentrated sulphuric acid also absorbs water. As there is no change of state, we call this type of substance *hygroscopic*.

Questions

1. Which one of the following compounds is a base?

A copper(II) nitrate
B iron(II) sulphate
C potassium chloride
D sodium hydroxide
E sulphur dioxide.

2. Which one of the following is the process by which two elements combine to form a single compound?

A combustion
B decomposition
C neutralization
D precipitation
E synthesis.

3. Which one of the following will dissolve in dilute nitric acid and also in aqueous sodium hydroxide?

A carbon monoxide
B copper(II) oxide
C iron(III) oxide
D magnesium oxide
E zinc oxide.

4. Which one of the following, in a $1 \, mol/dm^3$ solution, has the lowest pH value?

A ammonia
B ethanoic acid
C hydrogen chloride
D sodium hydroxide
E sugar.

5. Which one of the following is the best method for preparing copper(II) sulphate?

A adding sodium sulphate solution to copper(II) chloride solution
B adding copper(II) carbonate to dilute sulphuric acid
C adding copper to sodium sulphate
D adding copper to dilute sulphuric acid
E passing sulphur dioxide over hot copper powder.

6. The following table shows the pH of a number of aqueous solutions:

SOLUTION	A	B	C	D	E
pH	10	4	2	7	8

a
   (i) Which solution is the most acidic?
  (ii) Which solution is the most alkaline?
 (iii) Which solution is neutral?
 (iv) Which solution could be sugar solution?
  (v) Which solution could be vinegar?

b Which of the following *could* give a neutral solution on mixing?
   (i) A + B
  (ii) C + D
 (iii) B + D
 (iv) C + E
  (v) A + B + C + D + E

7. AQUEOUS AMMONIA, CITRIC ACID, COPPER, COPPER(II) CARBONATE, COPPER(II) OXIDE, HYDROCHLORIC ACID, MAGNESIUM, SODIUM HYDROXIDE.
Choose from the above list:

a an alkali
b a base which is not an alkali
c a substance that produces carbon dioxide gas when added to dilute sulphuric acid
d a mixture
e a strong acid
f a weak acid
g a substance which produces hydrogen gas when added to dilute sulphuric acid
h a substance that reacts with acids to form salt and water only
i a weak alkali
j a substance found in lemons

8. ACIDIC, ALKALINE, ANHYDROUS, HYDRATED, DELIQUESCENT, EFFLORESCENT, HYDROGEN, HYDROXIDE, INDICATOR, NEUTRAL, PRECIPITATE.
Choose from the above list the word needed to complete the following:

a When copper(II) nitrate crystals are left on a watch glass in the laboratory they turn to a blue solution. This is because copper nitrate crystals are _____ .
b Any solution with a pH less than 7 must be _____ .
c Any solution with a pH greater than 7 must be _____ .
d An acidic solution must contain an excess of (i) _____ ions over (ii) _____ ions.
e Salts which lose water on heating are said to be _____ .

f When washing soda crystals ($Na_2CO_310H_2O$) are left in an open laboratory they lose weight and become powdery. This is because washing soda crystals are _____ .

g A solution of sodium chloride in water is _____ .

h A substance that has different colours in acidic and alkaline solutions is known as an _____ .

i A solid formed on mixing two solutions is known as a _____ .

9. The following are methods or preparing salts:
   (i) metal + acid   (ii) alkali + acid
   (iii) base + acid   (iv) carbonate + acid
   (v) precipitation

   Which of these methods would you use to prepare the salts below?

   a copper(II) sulphate
   b magnesium sulphate
   c lead(II) chloride
   d sodium chloride
   e zinc nitrate
   f calcium carbonate
   g silver chloride
   h zinc chloride
   i potassium sulphate
   j lead(II) nitrate.

10. What two chemicals would you use to make each of the following salts?

    a copper(II) sulphate
    b magnesium chloride
    c lead(II) sulphate (insoluble)
    d sodium nitrate
    e copper(II) carbonate (insoluble)

11. Complete the following word equations and then, if you can, change them into balanced symbol equations:

    a zinc + hydrochloric acid →
    b lead(II) carbonate + nitric acid →
    c sodium hydroxide + sulphuric acid →
    d copper(II) oxide + hydrochloric acid →
    e silver nitrate + sodium chloride →
    f magnesium + sulphuric acid →
    g potassium carbonate + nitric acid →
    h lead(II) nitrate + sodium sulphate →

12. The waste water from a factory was found to be acidic. In a test, $5\,dm^3$ of the waste water needed 1 g of calcium hydroxide to neutralise it. The factory puts $100\,000\,dm^3$ of waste water into a local river every hour. This water *must* be neutral.

    a How much calcium hydroxide will be needed to neutralise the waste water every hour?
    b How much lime will be used each day? (Assume that the factory works 24 hours every day).

13. A solution containing $0.1\,mol/dm^3$ of hydrochloric acid has a pH of 1.
    A solution containing $0.1\,mol/dm^3$ of ethanoic acid has a pH of 3. What explanation can you suggest for this difference?

14. Electron transfer takes place when a metal reacts with an acid.

    a (i) Write the ionic equation for the reaction between zinc and dilute sulphuric acid.
       (ii) Why is the zinc said to have been oxidised in reaction **a**(i)?
    b Pure zinc reacts only slowly with dilute sulphuric acid; and calcium, although more reactive than zinc, reacts even more slowly.
       (i) Name a substance that can be added to a mixture of zinc and dilute sulphuric acid to speed up the rate of the reaction.
       (ii) Why is the reaction of calcium with dilute sulphuric acid so slow?
    c (i) Write the equation for the reaction between sodium and dilute ethanoic acid (acetic acid).
       (ii) Suggest why, although the reaction of sodium with dilute hydrochloric acid is almost explosive, the reaction between sodium and dilute ethanoic acid is only a little faster than that between sodium and water.
    d Dilute hydrochloric acid was added to an excess of magnesium ribbon. After a very short delay, effervescence became increasingly rapid, then slowed down and eventually stopped. Explain these observations.

    [C]

15.

    a Describe the observations, if any, you would make when dilute sulphuric acid reacts with:
       (i) zinc metal
       (ii) sodium hydroxide solution
       (iii) copper(II) carbonate
       (iv) blue litmus paper.

    Equations for the reactions are not required.
    b Giving full practical details, state clearly how you would obtain pure, dry crystals of copper(II) sulphate ($CuSO_4.5H_2O$) from a sample of the crystals contaminated with copper(II) carbonate.
    c Oxides of elements may be classified as acidic, basic, neutral or amphoteric.

    Give the *name* of one example of each of these types of oxide.   [C]

16. Lithium, sodium and potassium are in Group I of the Periodic Table. These three metals are known collectively as "alkali metals".
    Chlorine, bromine and iodine are in Group VII of the Periodic Table. These three non-metals are known collectively as "halogens". Halogen means "salt producer".
    Helium, neon and argon are in Group 0 of

the Periodic Table. These three gases are known collectively as "noble (inert) gases".

a   What is *an alkali*? Give the name and formula of *one* alkali.

b   What is a *salt*? Give the name and formula of a salt formed between an alkali metal and a halogen and describe how a pure sample of this salt could be prepared by a titration method.

c   Why are helium, neon and argon unreactive?

d   Write down, in each case, the formula of (i) the alkali metal ion, (ii) the halogen ion, that has the same arrangement of electrons as the neon atom.

e   142 g of chlorine gas and 168 g of krypton gas (Kr) occupy equal volumes under the same conditons of temperature and pressure. What can you deduce about the gas krypton from this information?

[C]

10 Electricity and matter

Electricity is a very convenient form of energy. We use it in our homes for lighting, heating, cooking, and for many other things. One of the big advantages of electricity is that it can easily be controlled. It can be turned on or off at the flick of a switch.

Electricity can be easily controlled because it passes through some materials but not through others.

A substance that allows electricity to pass through it is called a *conductor* of electricity.

A substance that does not allow electricity to pass through it is called a non-conductor or *insulator*.

10.1 Conductors and insulators

If you were given a selection of substances in the laboratory and asked to divide them into conductors and insulators it would be a fairly easy job.

Experiment 10.1 Insulators and conductors

You will need: copper foil, zinc foil, lead, graphite, roll sulphur, wood, plastic, paraffin wax and sugar.

Set up the circuit as shown in Fig 10.1 and find out if each substance is a conductor or insulator.

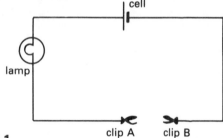

Fig 10.1

You will need: distilled water and solutions of copper(II) sulphate, sodium chloride, dilute hydrochloric acid, dilute sodium hydroxide and sugar.

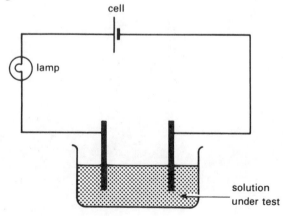

Fig 10.2

Set up the apparatus as shown in Fig 10.2 and find out which of the solutions conduct an electric current.

If the substance was a conductor the lamp would glow. If it was an insulator the lamp would not light up.

Table 1 shows some conductors and insulators that you might find in the laboratory.

CONDUCTORS	INSULATORS
Copper, aluminium, zinc, iron, steel, graphite, salt water, aqueous copper(II) sulphate, dilute hydrochloric acid, dilute sulphuric acid, aqueous sodium hydroxide	Sulphur, oxygen, iodine, carbon dioxide, ethanol, water, wax, polythene, PVC, copper(II) sulphate crystals, sodium chloride crystals, sugar, any pure acids, a solution of sugar in water

Table 1

Insulators

Looking at the insulators in Table 1 you will see they can be divided into a number of groups.

1. *Non-metal elements*: all non-metal elements except graphite are insulators. They will not conduct electricity in any state; solid, liquid or gas.

2. *Covalent compounds*: all covalent compounds are insulators. It does not matter whether they are solids like wax and sugar, liquids like ethanol and water, or gases like carbon dioxide. Pure acids fit into this group. So do the plastics, such as PVC, which are so widely used as insulators.

3. *Ionic solids*: ionic substances like sodium chloride and copper(II) sulphate are insulators, but only when they are solids.

Conductors

Looking at the conductors in Table 1 you will see they can also be divided up into a number of groups.

1. *Metals*: all metals conduct electricity when solid or molten. Some metals are better conductors than others.

Aluminium and copper are two of the best conductors. This is why they are so widely used for electrical wiring.

2. *Acidic solutions*: pure acids are covalent but when they dissolve in water they form ions and become conductors.

3. *Ionic compounds*: ionic compounds like sodium chloride conduct electricity when they are molten, or dissolved in water, but not when they are solid.

10.2 How do conductors conduct?

For a substance to conduct electricity, it must contain charged particles which are free to move. In metals (and graphite) the charged particles are electrons. The electrons that are free to move are the outer shell electrons or valency electrons. It is these valency electrons that carry an electric current through metals.

All other conductors contain ions that are free to move. It is these ions that carry the electric current through the substance. Ionic solids such as sodium chloride do not conduct electricity because the ions are firmly held and cannot move. Only when molten or dissolved in water are the ions free to move.

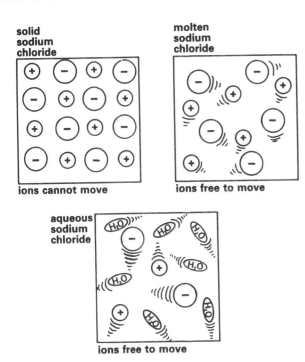

Fig 10.3

When electricity passes through a metal, the metal is chemically unaffected. However, when electricity passes through an ionic substance (either molten or in solution) the substance is broken down in some way.

A substance that conducts electricity and is decomposed by the electricity is known as an *electrolyte*.

10.3 Passing electricity through electrolytes

Fig 10.4 shows an apparatus suitable for passing electricity through an electrolyte.
When electricity is passed through an electrolyte, the electricity enters and leaves the electrolyte via electrical contacts. These contacts are known as *electrodes* and are usually made of graphite or platinum.

The positive electrode is known as the *anode*.
The negative electrode is known as the *cathode*.

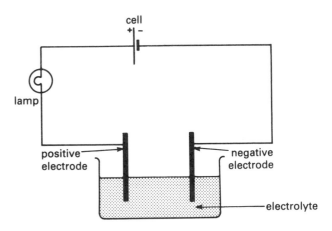

Fig 10.4

The ions in the electrolyte are attracted towards the electrodes.

Negative ions (called *anions*) are attracted towards the anode.
Positive ions (called *cations*) are attracted towards the cathode.

When electricity is passed through an electrolyte, chemical reactions take place at the electrodes, and the electrolyte is broken down. This process is known as electrolysis.

Electrolysis is the process in which a substance conducts electricity and is decomposed by it.

Many substances can be broken down or decomposed by heating.

Electrolysis is also a way of breaking down substances. It uses electrical energy instead of heat energy. Consider some of the ways in which electrolysis can be used.

1. Electrolysis of molten lead(II) bromide

Inert electrodes such as graphite or platinum are used.

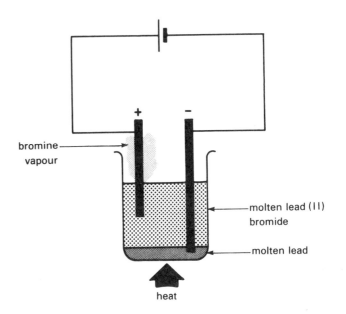

Fig 10.5

When electricity is passed through molten lead(II) bromide, it is broken down to form lead metal and bromine vapour:

lead(II) bromide → lead + bromine
$$PbBr_2 \rightarrow Pb + Br_2$$

The lead is formed at the cathode and the bromine is formed at the anode.

We can consider the reactions at the anode and at the cathode separately.

At the anode

Negative bromide ions are attracted towards the positive anode. At the anode they lose electrons and form bromine molecules:

bromide ions → bromine molecules + electrons
$$2Br^- \rightarrow Br_2 + 2e^-$$

At the cathode

Positive lead ions are attracted towards the negative cathode. At the cathode they gain electrons and form lead atoms:

lead ions + electrons → lead atoms
$$Pb^{2+} + 2e^- \rightarrow Pb$$

2. Electrolysis of copper(II) sulphate solution

The way in which copper(II) sulphate solution conducts electricity depends on the electrode material.

(i) With platinum electrodes

Copper metal is formed at the cathode and oxygen gas is formed at the anode.

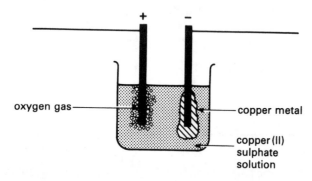

Fig 10.6

The solution contains copper(II) ions (Cu^{2+}) and sulphate ions (SO_4^{2-}) from the ionic copper(II) sulphate. It also contains some hydrogen ions (H^+) and hydroxide ions (OH^-) because water is slightly ionised.

At the anode

Hydroxide ions lose electrons forming water molecules and oxygen molecules:

hydroxide → water + oxygen + electrons
ions   molecules   molecules
$$4OH^- \rightarrow 2H_2O + O_2 + 4e^-$$

The sulphate ions are unchanged.

At the cathode

Copper(II) ions gain electrons to form copper atoms.

copper(II) ions + electrons → copper atoms
$$Cu^{2+} + 2e^- \rightarrow Cu$$

(ii) With copper electrodes

Copper is still formed on the cathode, but the reaction at the anode is different. Instead of oxygen gas being formed, the anode dissolves.

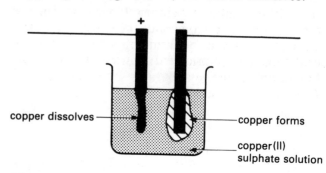

copper dissolves — — copper forms
— copper(II) sulphate solution

Fig 10.7

At the anode

Copper atoms from the anode lose electrons to form copper(II) ions. These ions pass into the solution.

copper atoms → copper(II) ions + electrons
$$Cu \rightarrow Cu^{2+} + 2e^-$$

The overall change is that copper is moved from the anode to the cathode. This makes the process suitable for copper plating. For copper plating, the object to be plated is made the cathode and a piece of pure copper is used as the anode.

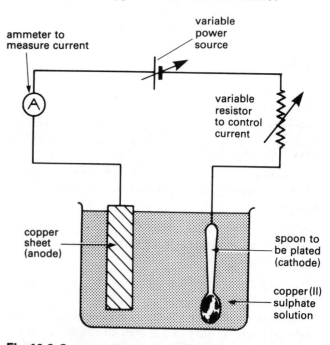

ammeter to measure current

variable power source

variable resistor to control current

copper sheet (anode)

spoon to be plated (cathode)

copper(II) sulphate solution

Fig 10.8 Copper plating a spoon

For successful copper plating the current, temperature and concentration of the electrolyte must be carefully controlled. The anode is sometimes arranged as a cylinder around the object to be plated. This gives a more even plating.

3. Electrolysis of dilute sulphuric acid

Pure water is a very poor conductor of electricity because it is a covalent compound. In pure water less than one molecule in one million is split into ions:

water $\rightleftharpoons$ hydrogen ions + hydroxide ions
$$H_2O \rightleftharpoons H^+ + OH^-$$

If a little sulphuric acid is added to water it becomes a good conductor of electricity. The electrolysis of water containing a little sulphuric acid (dilute sulphuric acid) can be shown in the laboratory using the apparatus shown in Fig 10.9. The electrodes used are graphite or platinum.

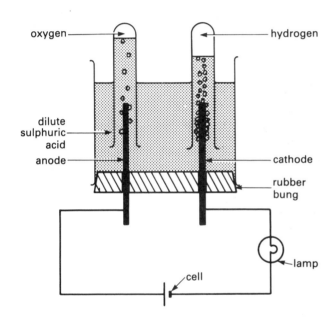

Fig 10.9 Electrolysis of acidified water

The electrolyte contains hydrogen ions (H^+) and sulphate ions (SO_4^{2-}) from the sulphuric acid. It also contains hydrogen ions (H^+) and hydroxide ions (OH^-) from the slightly ionised water.

At the anode

The hydroxide ions lose electrons forming water molecules and oxygen molecules:

hydroxide $\rightarrow$ water + oxygen + electrons
ions   molecules   molecules
$$4OH^- \rightarrow 2H_2O + O_2 + 4e^-$$

The sulphate ions are unchanged.

At the cathode

Hydrogen ions gain electrons to form hydrogen molecules:

hydrogen ions + electrons $\rightarrow$ hydrogen
                        molecules
$$2H^+ + 2e^- \rightarrow H_2$$

The overall change is that water is split up into its elements.

water $\rightarrow$ hydrogen + oxygen
$$2H_2O \rightarrow 2H_2 + O_2$$

This is a very convenient way of producing hydrogen gas. However, it is not used on a large scale in most parts of the world because of the high cost of electricity.

4. The electrolysis of concentrated hydrochloric acid

The electrolyte contains hydrogen ions (H^+) and chloride ions (Cl^-) from the hydrochloric acid. It also contains hydrogen ions (H^+) and hydroxide ions (OH^-) from the slightly ionised water.

At the anode

The chloride ions lose electrons forming chlorine molecules.
$$2Cl^- \rightarrow Cl_2 + 2e^-$$

At the cathode

Hydrogen ions gain electrons to form hydrogen molecules.
$$2H^+ + 2e^- \rightarrow H_2$$

5. Electrolysis of a concentrated solution of sodium chloride

The electrolyte contains sodium ions (Na^+) and chloride ions (Cl^-) from the sodium chloride. It also contains hydrogen ions (H^+) and hydroxide ions (OH^-) from the slightly ionised water.

At the anode (graphite or platinum)

The chloride ions lose electrons forming chlorine molecules
$$2Cl^- \rightarrow Cl_2 + 2e^-$$

At the cathode (graphite or platinum)

Hydrogen ions gain electrons to form hydrogen molecules.
$$2H^+ + 2e^- \rightarrow H_2$$

Quantity of electricity

Electrical charge is measured in coulombs (C).

One coulomb of charge = 1 ampere of current flowing for 1 second.

Number of coulombs = number of amperes × time in seconds

The passage of one mole of electrons (L electrons) is the same as 96500 coulombs of charge. (The Faraday constant = 96500 C/mol.)

Calculations on electrolysis

1. What mass of silver is deposited by 1 mole of electrons?

Write the equation:
$$Ag^+ + e^- \rightarrow Ag$$
1 mole of electrons gives 1 mole of silver atoms. The mass of one mole of silver is 108 grams.
Answer = 108 grams.

67

2. What mass of copper is deposited by 1 ampere of current passing for 32 minutes 10 seconds?

Convert the time to seconds

32 minutes 10 seconds = 1930 seconds

Number of coulombs = amperes × seconds

$$= 1 \times 1930$$
$$= 1930 \text{ coulombs}$$

Write the equation:

$$Cu^{2+} + 2e^- \rightarrow Cu$$

2 moles of electrons gives 1 mole of copper atoms.

193000 coulombs gives 64 grams of copper.

1930 coulombs gives 0·64 g.

3. Calculate the charge on an aluminium ion if 5·4 grams of aluminium is deposited by a current of 5 amperes passing for 3 hours 13 minutes.

3 hours 13 minutes = 193 minutes

$$= 11580 \text{ seconds}$$

Coulombs passed = 5 × 11580

$$= 57900 \text{ coulombs}$$

5·4 grams of aluminium is discharged by 57900 coulombs.

One mole of aluminium atoms has a mass of 27 grams

27 grams of aluminium is discharged by

$$\frac{57900 \times 27}{5 \cdot 4}$$

$$= 289500 \text{ coulombs}$$

$$= \frac{289500}{96500}$$

$$= 3 \text{ moles of electrons}$$

∴ Charge on the aluminium ions is 3+

(Positive ions because aluminium is a metal.)

Note: the charge on ions must be a whole number.

4. What volume of oxygen is produced at r.t.p. when a current of 2 amperes is passed for 6 minutes 26 seconds through a solution containing hydroxide ions?

$$4OH^- \rightarrow 2H_2O + O_2 + 4e^-$$

4 moles of electrons produces 24000 cm³ of oxygen.

386000 coulombs produces 24000 cm³ of oxygen

Coulombs used = 2 × 386

$$= 772 \text{ coulombs}$$

772 coulombs will produce $\dfrac{24000 \times 772}{38600}$ cm³

$$= 48 \text{ cm}^3$$

10.4 Making use of electrolysis

Electrolysis is of benefit to man in a number of ways:

1. It has allowed him to produce large quantities of the more reactive metals such as aluminium. Before the discovery of electricity aluminium was a very rare and precious metal. Many aluminium compounds existed, but it was difficult to extract aluminium from them. Nowadays, vast quantities of aluminium are made by the electrolysis of molten aluminium oxide (see Chapter 11). It is used in so many ways that hardly a day goes by without us using the metal in some way or other. Think how many ways you have used aluminium today.

2. Electrolysis allows us to purify some metals. Copper can be purified (refined) in this way.

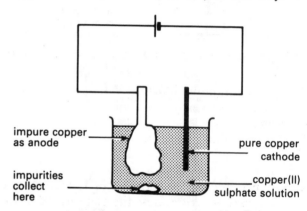

Fig 10.10 Purifying copper

When the current flows copper is dissolved from the impure anode and transferred to the pure cathode. The impurities are left behind.

3. The discovery of electrolysis allowed the modern electroplating industry to develop. Metals like steel are often plated. Car bumpers are often chromium plated because it makes them look more attractive and it prevents the steel underneath from rusting. By tin plating steel cans, a container seeming to be made of pure unreactive tin is produced. A can made of pure tin would be very much more expensive and certainly not as strong.

4. Electrolysis of brine (aqueous sodium chloride) produces chlorine, hydrogen and sodium hydroxide. In a diaphragm cell chlorine is produced at a titanium anode and hydrogen at a steel cathode. To prevent the electrode products from mixing, a

Name of substance	Electrolyte	Other substances formed at the same time
aluminium	molten alumina in cryolite	oxygen
sodium hydroxide	sodium chloride solution	hydrogen and chlorine
sodium	molten sodium chloride	chlorine

Table 2

porous sheet of asbestos separates the anode and cathode compartments. The resulting solution, with its hydrogen ions and chloride ions removed, is one of sodium hydroxide.

Table 2 shows some substances manufactured by electrolysis.

Questions

1. When the experiment shown in the diagram was set up, the electric bulb lit, but there were no products at the electrodes.

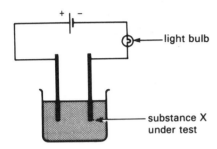

Which one of the following is most likely to be X?

A aqueous sodium chloride
B a solution of sugar in water
C bromine
D molten sodium chloride
E mercury.

2. When dilute hydrochloric acid is electrolysed using carbon electrodes, the gas formed at the cathode (negative electrode) is:

A carbon dioxide
B chlorine
C hydrogen
D hydrogen chloride
E oxygen.

3. Which one of the following processes occurs at the electrodes when molten lead(II) bromide is electrolysed?

	positive electrode	negative electrode
A	oxidation	reduction
B	no reaction	no reaction
C	no reaction	oxidation
D	reduction	no reaction
E	reduction	oxidation

4. Which one of the following graphs represents how the mass of the copper anode (positive electrode) varies with time, when an electric current is passed through aqueous copper(II) sulphate using copper electrodes?

5. When a current of 1 A (ampere) is passed through molten lead(II) chloride for 200 minutes the volume of chlorine produced, measured at room temperature and pressure is:

A $0.5\,dm^3$
B $1.0\,dm^3$
C $1.5\,dm^3$
D $10\,dm^3$
E $15\,dm^3$.

(The Faraday constant = 96000 C(coulombs)/mol; 1 mol of chlorine gas has a volume of $24\,dm^3$ at room temperature and pressure.)

6. ALUMINIUM, GRAPHITE, OXYGEN, SALT, SUGAR, SULPHUR, ZINC.
Choose from the above list:

a two elements which conduct electricity at room temperature.
b two elements which do not conduct electricity at room temperature.
c an element extracted from its ore by electrolysis.
d a substance which conducts electricity when molten but not when solid.
e a substance which conducts electricity when solid or molten.
f a non metal which conducts electricity when solid.
g A substance made up of ions.
h A compound made up of molecules.

7.

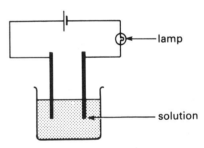

In turn, four different solutions were placed in the circuit shown in the diagram.
The solutions were: (i) copper(II) sulphate solution; (ii) sodium chloride solution; (iii) sugar solution; (iv) dilute sulphuric acid.

Decide which solution was in the circuit when the following results were obtained. Give reasons for your choice:

a The lamp lights; a colourless gas is formed at both electrodes.

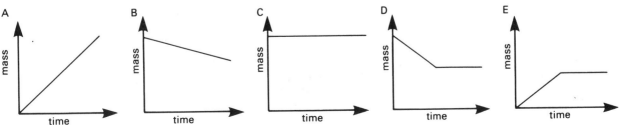

b The lamp lights; a pinky brown solid is formed on the negative electrode.

c The lamp lights; a green gas is formed at the positive electrode.

d The lamp does not light.

8. A small leaf was painted with a graphite paste and then made the cathode in the circuit shown in the diagram. The leaf became coated with copper metal.

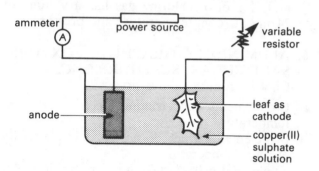

a Why must the leaf be painted with graphite paste before it is plated?

b Why is the leaf made the cathode rather than the anode?

c What material is suitable for the anode?

d What is the purpose of the ammeter?

e What is the purpose of the variable resistor?

f Why cannot an alternating current be used to copper plate the leaf?

9. The following results were obtained during the electrolysis of $500 \, cm^3$ of aqueous copper(II) sulphate using carbon electrodes.

Total mass of copper deposited on the cathode/g	Number of coulombs passed
0·33	1000
1·00	3000
1·65	5000
2·30	7000
3·00	9000
3·00	11000
3·00	13000

a Draw a labelled diagram of the apparatus you would use, including the electrical circuit, in order to carry out this electrolysis.

b Plot the experimental results as a graph.

c Write an ionic equation, including state symbols, to represent the reaction at the cathode during the electrolysis.

d Explain the shape of your graph obtained in **b**.

e Calculate the concentration, in mol/dm^3, of the original aqueous copper(II) sulphate.

[C]

10.

a *Name* the types of particles present in the crystal lattices of: (i) copper metal (ii) sulphur (iii) sodium chloride.

b Explain why, in the solid state, copper can

conduct electricity but sulphur and sodium chloride cannot.

c A metal spoon is to be electroplated with copper. Give a diagram of the plating cell and the circuit required, clearly labelling the materials to be used for the anode, cathode and electrolyte.

d Calculate the current required to deposit 0·16 g of copper in one minute.

[C]

11.

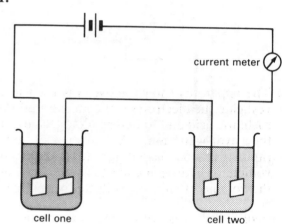

Cell One has platinum electrodes dipping into dilute sulphuric acid.

Cell Two has platinum electrodes dipping into copper(II) sulphate solution.

a By reference to the diagram, copy and then complete the table below.

		Name of material liberated	Equation for the electrode reaction
Cell One	Positive electrode		
	Negative electrode		
Cell Two	Positive electrode		
	Negative electrode		

b What change, if any, would you expect to see in the reading of the current meter if, in separate experiments,

(i) the solution in Cell Two were replaced by a solution of sugar.

(ii) a few drops of barium hydroxide solution were added to the solution in Cell One? Give brief reasons for your answers.

c Name *one* material which is manufactured by electrolysis. State the nature of the electrodes and of the electrolyte used in the manufacturing process you choose.

[C]

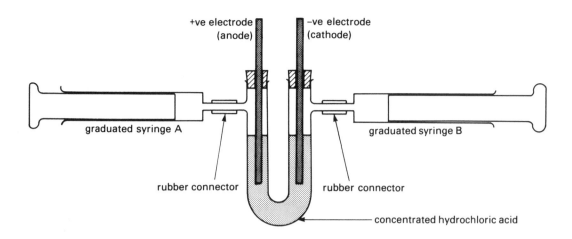

+ve electrode (anode)

−ve electrode (cathode)

graduated syringe A

graduated syringe B

rubber connector

rubber connector

concentrated hydrochloric acid

12. The apparatus shown above was used to investigate the electrolysis of concentrated hydrochloric acid and to determine the charge on the hydrogen ion. A current of 0·6 A (amperes) was passed for 16 minutes; the volume of hydrogen collected was 72 cm³ and the volume of chlorine collected was 60 cm³.

a *Name* the gas collecting in syringe A. Describe *one* chemical test for this gas.

b Why are the electrodes made of carbon and not of a metal such as iron?

c Explain why the volume of chlorine collected is less than the volume of hydrogen.

d Use the numerical information given to calculate the number of unit charges carried by a hydrogen ion. (Assume that 1 mole of hydrogen gas occupies 24 dm³ under the conditions of the experiment and that the Faraday constant is 96000 coulombs per mole of electrons.)

e The experiment was repeated using dilute sulphuric acid and platinum electrodes, instead of concentrated hydrochloric acid and carbon electrodes. The same current was passed for the same length of time.
  (i) What products are set free at the electrodes?
  (ii) What volumes of gases are obtained?
  [C]

13.

a Sodium hydroxide is manufactured from brine on a large scale, using electrolytic methods. Describe *one* such method. In your account, you should draw a simple labelled diagram of the cell, state the materials of which the electrodes are made, write ionic equations for the reactions taking place at the electrodes and list any by-products which are obtained.

b For *two* of the by-products which are obtained, give two uses of each and say how the materials so obtained have benefited society.
  [L]

14.

a Describe, with the aid of a labelled diagram, the electrolysis of an aqueous solution of copper(II) sulphate using carbon electrodes. Your answer should indicate how the experiment would be set up, observations you would make during the electrolysis and reactions taking place at the electrodes.

b If both electrodes had been made of copper, describe any differences in observations and electrode reactions that you would note.

c Briefly explain the industrial importance of *one* of these experiments.
  [L]

15.

a Consider the following substances: lead(II) nitrate, sugar, hydrogen chloride, sodium, calcium hydroxide, ethanol, potassium iodide, sulphur.

  Name those substances that:
  (i) conduct electricity in the solid state
  (ii) conduct electricity *both* when molten *and* in aqueous solution
  (iii) do not conduct electricity themselves but form conducting liquids when dissolved in water
  (iv) do not readily conduct electricity under any of these conditions.
  You may list any substance more than once if appropriate.

b An aqueous solution of calcium hydroxide is electrolysed between carbon electrodes.
  (i) What gas would you expect to be produced at the anode (positive electrode)? It is observed that, during the electrolysis, the mass of the anode decreases and the solution round it becomes milky.
  (ii) Suggest an explanation for these observations.

c A strip of moistened filter paper is laid on a microscope slide. A drop of silver nitrate solution is placed near one end of the paper and a drop of potassium iodide solution near the other end. Using inert electrodes, the apparatus was connected to a suitable d.c. supply. After some time, a pale yellow streak appeared as shown in the diagram.

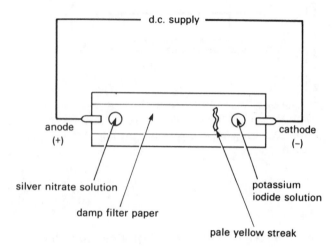

(i) Give the formulae of the ions present in the solutions of silver nitrate and potassium iodide.
(ii) Explain the process leading to the formation of the streak and explain why the streak appears nearer the cathode than the anode.

[C]

16. A pure specimen of molten lead(II) bromide was electrolyzed in a suitable apparatus using inert electrodes.

a Give the formulae of the ions present in the liquid.
b What changes would you expect to SEE at each of the electrodes?
c Write ionic equations to show the changes taking place at each of the electrodes.
d In such an experiment, a current of 0·2 amperes was passed through the molten lead(II) bromide for four minutes.

(i) What quantity of electricity passed?
(ii) What would be the mass of the product liberated at the negative electrode?
(iii) Assuming all the product liberated at the positive electrode to be in the form of a gas, what volume would it occupy at room temperature and atmospheric pressure?

[L]

11 The metal elements

When we first think of metals we may think of solids that conduct electricity, are strong and can be hammered into shape without shattering. In Chapter 6 we saw that to a chemist the word metal means much more than this. Metals have many properties in common. These common properties distinguish them from non-metals.

Even though metals have many common properties they are different in many ways. *Example*: Potassium bursts into flames as soon as it is dropped into water, whereas gold can be recovered untarnished from the sea bed after hundreds of years. Chemists say that potassium is a very *reactive* metal and gold is a very *unreactive* metal.

11.1 The reactivity of metals

It is useful to arrange metal elements in a reactivity series. This is a list of elements in which the most reactive element is at the top and the metals become less reactive as we go down the list. A reactivity series of some common metals is shown in the following table.

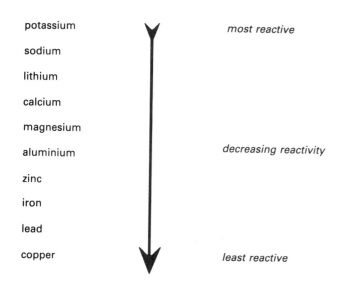

potassium — most reactive
sodium
lithium
calcium
magnesium
aluminium — decreasing reactivity
zinc
iron
lead
copper — least reactive

Some simple experiments can be performed in the laboratory to show that the above reactivity list is correct.

A. The reactions of metals

1. With air

Nearly all metals react with oxygen in the air to form an oxide. This process is speeded up by heating. In some cases, the reaction is quite vigorous. The following metals will ignite on heating, and an exothermic reaction occurs. Heat and light energy are given out:

potassium —lilac flame
sodium   —yellow flame
calcium   —red flame
magnesium—white flame

Clearly, these are very reactive metals

If some aluminium is heated, nothing seems to happen. This is because there is already an oxide film on the surface of the metal. This prevents further reaction. If the oxide layer is removed by treating with aqueous mercury(II) chloride, the clean aluminium surface reacts rapidly in air. A white oxide layer can be seen forming and the metal gets quite hot.

Both zinc dust and iron filings glow when they are heated in air but they do not catch fire. They are less reactive than aluminium.

Lead and copper both change colour on heating as oxide films are formed.

All the above reactions can be represented by the simple word equation

metal + oxygen → metal oxide

2. With water

Table 1 overleaf shows what happens when a *small* piece of each metal is dropped into a trough of cold water.

It can be seen that as we go down the reactivity series metals react less violently with water. Metals below magnesium in the reactivity series do not react with cold water.

3. Reaction of metals with steam

To investigate the reaction of metals with steam the apparatus in Fig 11.1 can be used:

The metal is strongly heated until it is very hot and then the mineral wool is heated so that steam passes over the hot metal. Table 2 overleaf shows what happens.

Note: Great care must be taken with this experiment to stop the water sucking back from the trough into the hot test tube.

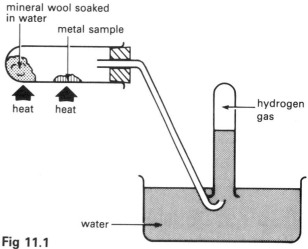

mineral wool soaked in water
metal sample
heat   heat
hydrogen gas
water

Fig 11.1

METAL	REACTION	EQUATION					ORDER IN OUR REACTIVITY SERIES
potassium	It reacts violently on the surface of the water. Enough heat is produced in the reaction to melt the potassium into a ball and to light the hydrogen gas formed. The hydrogen burns with a lilac flame. The remaining solution is alkaline.	potassium 2K	+ water + $2H_2O$	$\rightarrow$ $\rightarrow$	potassium hydroxide 2KOH	+ hydrogen + H_2	first
sodium	It reacts violently on the surface of the water. Enough heat is produced to melt the sodium, but not enough to ignite the hydrogen gas formed. The remaining solution is alkaline.	sodium 2Na	+ water + $2H_2O$	$\rightarrow$ $\rightarrow$	sodium hydroxide 2NaOH	+ hydrogen + H_2	second
lithium	The lithium floats on the surface of the water reacting vigorously to produce hydrogen gas and an alkaline solution. The reaction is not as violent as that of sodium. Not enough heat is produced to ignite the hydrogen gas.	lithium 2Li	+ water + $2H_2O$	$\rightarrow$ $\rightarrow$	lithium hydroxide 2LiOH	+ hydrogen + H_2	third
calcium	Calcium sinks and reacts steadily producing hydrogen gas and leaving an alkaline solution.	calcium Ca	+ water + $2H_2O$	$\rightarrow$ $\rightarrow$	calcium hydroxide $Ca(OH)_2$	+ hydrogen + H_2	fourth
magnesium	Magnesium turnings sink and a very slow reaction takes place, producing hydrogen gas.	magnesium Mg	+ water + $2H_2O$	$\rightarrow$ $\rightarrow$	magnesium hydroxide $Mg(OH)_2$	+ hydrogen + H_2	fifth

The apparatus needs to be left for several days to collect a single test tube of hydrogen gas. The solution left is slightly alkaline.

- hydrogen gas
- water
- magnesium turnings

Table 1

METAL	REACTION	EQUATION					ORDER OF REACTIVITY
magnesium	The hot magnesium reacts very violently with the steam. A bright white glow is produced and hydrogen gas is collected. White, powdery magnesium oxide is left in the tube.	magnesium Mg	+ steam + H_2O	$\rightarrow$ $\rightarrow$	magnesium oxide MgO	+ hydrogen + H_2	fifth
zinc	When steam is passed over red hot zinc powder the zinc glows slightly more brightly and hydrogen gas is collected. Zinc oxide powder is left in the tube. It is yellow when hot and white when cold.	zinc Zn	+ steam + H_2O	$\rightarrow$ $\rightarrow$	zinc oxide ZnO	+ hydrogen + H_2	sixth
iron	When steam is passed over red hot iron some hydrogen is formed. The iron must be constantly heated while the steam is passed over it.	iron 3Fe	+ steam + $4H_2O$	$\rightleftharpoons$ $\rightleftharpoons$	tri-iron tetraoxide Fe_3O_4	+ hydrogen + $4H_2$	seventh

Table 2

As we go down the reactivity series metals react less violently with steam. Metals below iron in this reactivity series do not react with steam.

4. Reaction of metals with dilute hydrochloric acid

Potassium, sodium, lithium and calcium (the most reactive metals) should *never* be added to dilute acids as they react explosively.

Magnesium, zinc, iron and lead, from the middle of the reactivity series, react with dilute hydrochloric acid to form the metal chloride and hydrogen gas.

Example:

magnesium + hydrochloric → magnesium + hydrogen
             acid   chloride

$$Mg(s) + 2HCl(aq) \rightarrow MgCl_2(aq) + H_2(g)$$

The reaction of dilute hydrochloric acid on magnesium is fairly fast, but the *lower* the reactivity of the metal the slower the reaction becomes.

Lead reacts very, very slowly with dilute hydrochloric acid.

Copper, (the least reactive metal in our list) does not react with dilute hydrochloric acid.

Similar reactions occur with dilute sulphuric acid.

The reactions of the metals are summarised in Table 3.

Note: in the reactions of metals with water, steam, or dilute hydrochloric acid, the metal is always changed into an ionic compound.

Therefore metal atoms change into positively charged metal ions:

Example:

$Mg \rightarrow Mg^{2+} + 2e^-$
metal atom → metal ion + electrons
$Na \rightarrow Na^+ + e^-$
metal atom → metal ion + electron
(In both cases the metal atom is being oxidised to the metal ion.)

The reactivity of a metal depends on the ease with which it forms positive ions. The more reactive metals like sodium form positive ions far more easily than the less reactive metals such as copper.

B. Displacement of metals from solutions of salts

If we look at part of the reactivity table of metals, we can make predictions about their reactions.

   Mg
   Al
   Zn
   Fe
   Pb
   Cu

We would expect magnesium to displace iron from aqueous iron(II) sulphate, and that iron would displace copper from aqueous copper(II) sulphate because magnesium forms ions more readily than iron and iron forms ions more readily than copper.

On the other hand, we would expect that lead will not react with aqueous zinc sulphate or with aqueous magnesium sulphate.

A more reactive metal will displace a less reactive metal from a solution of one of its salts.

Example: If a small piece of zinc is dropped into blue copper(II) sulphate solution, brown copper metal is formed as the zinc reacts.

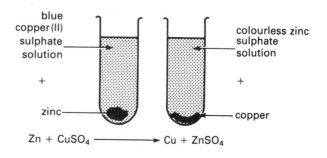

$$Zn + CuSO_4 \longrightarrow Cu + ZnSO_4$$

Fig 11.2

The more reactive metal (zinc) has displaced the less reactive metal (copper) from a solution of one of its salts (copper(II) sulphate).

Note: if copper is placed in zinc sulphate solution no reaction takes place.

METAL	REACTION WITH OXYGEN	REACTION WITH WATER	REACTION WITH DILUTE HYDROCHLORIC ACID
potassium sodium	React to give oxide when heated	React with cold water	Explosive
lithium calcium magnesium	↓ Reaction becomes less violent		React forming metal chloride and hydrogen
aluminium zinc		Protected by oxide film	
iron		Only react with steam	
lead		No reaction	Very slow
copper			No reaction

Table 3

In the displacement reactions of metals the more reactive metal changes into metal ions:

Example: $Zn \rightarrow Zn^{2+} + 2e^-$ (zinc atoms are oxidised)

The ions of the less reactive metal are changed into metal atoms.

Example: $Cu^{2+} + 2e^- \rightarrow Cu$ (copper ions reduced)

The reaction between zinc and aqueous copper(II) sulphate is:

$Zn(s) + Cu^{2+}(aq) \rightarrow Zn^{2+}(aq) + Cu(s)$

Why is this reaction a redox reaction?

You might think that sodium would displace almost all metals from solutions of their salts, but as the metal reacts rapidly with water no displacement reaction is seen.

Experiment 11.1. Displacing metals from solutions

You will need six pieces of foil (5 mm × 5 mm) of each of the following metals: aluminium, copper, lead, magnesium and zinc, together with six clean iron nails.

You will also need aqueous solutions of the nitrates of the following metals: aluminium, copper, lead, magnesium and zinc, together with aqueous ammonium iron(II) sulphate.

Place a 1 cm depth of each solution in separate test tubes. To each tube, add a small piece of aluminium foil. If *any* change is observed (for example, the colour of the solution changes or a precipitate appears on the metal), it means the metal is displaced from solution. If this occurs put a tick in a copy of the table below.

After five minutes empty all the tubes, rinse them and repeat the experiment using copper foil. Keep on repeating the experiment until you have used each metal in all of the solutions.

SOLUTION \ METAL	Al	Cu	Pb	Mg	Zn	Fe
Al^{3+}(aq)						
Cu^{2+}(aq)						
Pb^{2+}(aq)						
Mg^{2+}(aq)						
Zn^{2+}(aq)						
Fe^{2+}(aq)						

Which column has the most ticks? This tells you which is the most reactive metal.

Which columns have the least ticks?

How can you explain the results obtained for the experiments using aluminium foil?

Write down the order of reactivity of the metals that you have used.

Does it agree with the reactivity table at the beginning of the chapter?

Write chemical and ionic equations for the reactions that have occurred.

Experiment 11.2 Displacing metals from their oxides

You will need: iron filings, zinc oxide, copper(II) oxide, lead(II) oxide, and some strips of ceramic paper.

Thoroughly mix approximately equal volumes of iron filings and lead(II) oxide. Heat the mixture on a strip of ceramic paper. A glow will spread through the mixture if a reaction occurs.

Repeat the experiment twice more, mixing the iron filings with each of the other two oxides.

Do your results agree with the reactivity series of metals?

Write equations for any reactions observed.

C. The action of heat on carbonates, hydroxides and nitrates

The stability of carbonates, hydroxides and nitrates depends on the position of the metal in the reactivity series.

The carbonates of sodium and potassium are unaffected by heat. Those of less reactive metals break down to give the metal oxide and carbon dioxide:

$\text{calcium carbonate} \rightarrow \text{calcium oxide} + \text{carbon dioxide}$

$CaCO_3 \rightarrow CaO + CO_2$

The hydroxides of potassium and sodium are also unaffected by heat. When those of less reactive metals are heated, the metal oxide and water are produced:

$\text{calcium hydroxide} \rightarrow \text{calcium oxide} + \text{water}$

$Ca(OH)_2 \rightarrow CaO + H_2O$

Potassium nitrate breaks down on heating to give potassium nitrite and oxygen:

$\text{potassium nitrate} \rightarrow \text{potassium nitrite} + \text{oxygen}$

$2KNO_3 \rightarrow 2KNO_2 + O_2$

Nitrates of the less reactive metals, such as lead and copper, break down to give the metal oxide, nitrogen dioxide and oxygen.

$\text{lead(II) nitrate} \rightarrow \text{lead(II) oxide} + \text{nitrogen dioxide} + \text{oxygen}$

$2Pb(NO_3)_2 \rightarrow 2PbO + 4NO_2 + O_2$

All these reactions are summarised in Table 4.

D. Obtaining metals from their oxides

We have already seen in this Chapter, that some metals readily combine with oxygen. If an element has a great attraction for oxygen, we might expect that it would be difficult to separate the metal from the oxide.

We are going to look at the reaction of metal

METAL	HEAT ON CARBONATES	HEAT ON HYDROXIDES	HEAT ON NITRATES
potassium sodium	No reaction	No reaction	Decompose producing oxygen as the only gas
lithium calcium magnesium aluminium zinc iron lead copper	Decompose producing the oxide and carbon dioxide	Decompose producing the oxide and water	Decompose producing the oxide, nitrogen dioxide and oxygen

Table 4

oxides with substances that also have an attraction for oxygen.

1. *Hydrogen*

The apparatus shown can be used to pass hydrogen over a heated metal oxide.

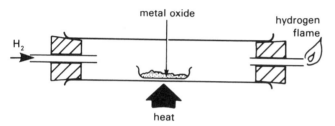

Fig 11.3

Copper(II) oxide reacts readily with hydrogen and the black powder changes to the pink-brown colour of copper.

copper(II) + hydrogen → copper + water
  oxide   (steam)
$$CuO(s) + H_2(g) → Cu(s) + H_2O(g)$$

Lead(II) oxide can be reduced in a similar manner, the yellow powder changing to silvery beads of molten lead.

The reaction with iron oxides is reversible, but reduction will occur.

$$Fe_3O_4(s) + 4H_2(g) \rightleftharpoons 3Fe(s) + 4H_2O(g)$$

Oxides of metals above iron in the reactivity series cannot normally be reduced using hydrogen.

2. *Carbon monoxide*

This gas can be used in the same apparatus as is used for reduction by hydrogen. Carbon monoxide reduces oxides of all metals below zinc in the reactivity series.

When a reaction occurs the metal and carbon dioxide are produced.

$$ZnO + CO → Zn + CO_2$$

3. *Carbon*

If lead(II) oxide is mixed with carbon and the mixture is heated, a red glow is seen and silvery beads of lead are formed.

lead(II) + carbon → lead + carbon
  oxide   dioxide
$$2PbO + C → 2Pb + CO_2$$

Like carbon monoxide, carbon cannot be used to reduce oxides of metals above zinc in the reactivity series.

11.2 The extraction of metals

The position of a metal in the reactivity series tells us what process can be used in its manufacture.

The three reducing agents we have just considered, cannot be used to reduce oxides of sodium or aluminium because these metals have a very strong attraction for oxygen. Reduction involves conversion of metal ions in the oxides to metal atoms. This involves addition of electrons.

To obtain the metals from compounds of sodium and aluminium we must use a powerful reducing agent. What better than electrons themselves—an electric current! Both these metals are extracted by electrolysis.

1. Sodium

The most commonly occurring sodium compound is sodium chloride (salt). Electrolysis of the molten compound produces chlorine at the positive electrode (anode) which is made of graphite

$$2Cl^- → Cl_2 + 2e^-$$

and molten sodium is formed at the negative electrode (cathode) which is made of iron.

$$Na^+ + e^- → Na$$

Sodium ions are reduced to sodium atoms by gaining electrons at the cathode.

2. Aluminium

Aluminium is obtained from an ore called bauxite, which is a hydrated form of aluminium oxide. After purification, the oxide is mixed with cryolite (Na_3AlF_6). The molten mixture is electrolysed. Molten aluminium is formed at the negative electrode and oxygen at the positive electrode. Both electrodes are made of graphite.

$$Al^{3+} + 3e^- → Al$$
$$2O^{2-} → O_2 + 4e^-$$

Anodised aluminium

We have already said that there is a thin film of aluminium oxide firmly held on the surface of the metal. It is possible to increase the thickness of this film—the process is called *anodising*.

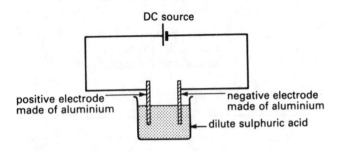

Fig 11.4

As a current is passed, the oxide film on the positive aluminium electrode becomes thicker. The thicker oxide layer is porous and can absorb dyes. The metal can therefore be given an attractive colour. Since the colour is *in* the metal itself the colours are long lasting. Aluminium that has been anodised and dyed is widely used for cheap jewellery and other decorative uses.

3. The extraction of zinc and iron

Reduction of oxides of zinc and iron can be done using the cheaper and less powerful reducing agent, carbon.

Coke is a relatively cheap source of fairly pure carbon. This itself can reduce oxides, but most of the reduction is probably a result of the carbon monoxide which is formed in the process.

Zinc

Zinc occurs naturally as its sulphide (ZnS). As the sulphide cannot be easily reduced, it is roasted in air to form zinc oxide. Sulphur dioxide is the other product of the reaction:

zinc + oxygen → zinc + sulphur
sulphide   oxide dioxide

Zinc oxide is then reduced in a blast furnace with coke. Hot air is blown in which reacts with the coke to form carbon monoxide. This reduces the oxide.

$$ZnO + CO \rightarrow Zn + CO_2$$

Iron

The importance of iron and steel to our lives can hardly be overestimated. Imagine the effect on a normal day if all the iron and steel suddenly disappeared: no cutlery, no cooker, no kettle, no water taps, no cars, no lorries, no trains. When we think of life without iron and steel we begin to realise how valuable iron is.

Fortunately, iron is the second most plentiful metal in the earth's crust. The main ores are haematite (Fe_2O_3) and magnetite (Fe_3O_4). It is from these ores that iron is extracted.

Iron is extracted from its ores as follows:
1. The iron ore is strongly heated in air. This removes moisture and some impurities.
2. The heated iron ore is fed into the top of a *blast furnace*, together with limestone and coke.
3. Hot air is blown into the furnace near the bottom. This blast of air allows coke to burn, making carbon dioxide and producing a very high temperature:

coke + air → carbon dioxide
C + O_2 → CO_2

4. The carbon dioxide reacts with hot coke to form carbon monoxide:

coke + carbon dioxide → carbon monoxide
C + CO_2 →   2CO

5. The carbon monoxide then reduces the *hot* iron ore to iron:

haematite + carbon → iron + carbon
            monoxide   dioxide
$$Fe_2O_3 + 3CO \rightarrow 2Fe + 3CO_2$$

6. The iron formed is molten and runs to the bottom of the furnace.

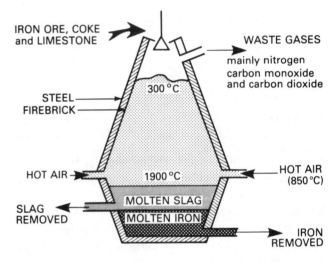

Fig 11.5 A blast furnace

The iron formed in the blast furnace would be extremely impure if limestone *was not* used. The added limestone breaks down to form calcium oxide in the blast furnace because of the high temperature.

limestone → calcium oxide + carbon dioxide
$$CaCO_3 \rightarrow CaO + CO_2$$

The calcium oxide reacts with silica (sand) impurities in the iron ore to form a molten slag:

calcium oxide + silica → slag (calcium silicate)
CaO   + SiO_2 → $CaSiO_3$

The slag does not mix with the iron but floats on top of it (like oil on water). Many of the impurities which would be in the iron, dissolve instead in the molten slag. The iron and slag are run off separately.

The iron run off from a blast furnace is known as pig iron, or cast iron. This iron contains 2–4%

carbon as well as other impurities such as sulphur, phosphorus and silicon. Cast iron is brittle and not strong. Most of the iron is changed into steel.

Steel making (Basic Oxygen Process)

Molten iron is poured into a large container. Oxygen at high pressure is then blown onto the surface of the molten iron. The blast of oxygen stirs the metal, and oxidises the impurities in the iron. The oxidised impurities either escape as gases or form a slag (with calcium oxide which is added). This can be poured off. The molten metal left is pure iron.

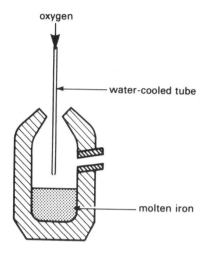

Fig 11.6

Carbon and/or certain metals are now stirred into the molten iron in the correct amounts to make the alloys required by various customers. The most common steel contains up to 0·5% carbon. It has many, many uses including car bodies, girders, screws and nails. Thousands of different types of steel have been made, all having slightly different properties. If the composition of a type of steel is changed slightly its properties may change greatly. This means that the steel industry must at all times carefully control its processes, if it is to meet the needs of its customers.

Experiment 11.3 Experiments on iron and its compounds

Place about 2 g of iron filings in a small flask containing about 20 cm³ of dilute sulphuric acid. Boil the mixture gently for about 5 minutes with a filter funnel placed in the neck of the flask to prevent acid spray escaping. **Wear your safety glasses.**

Allow the mixture to cool a little and filter off the unreacted iron. What colour is the resulting solution of iron(II) sulphate? Divide the filtrate into two portions.

1. To one, add about 2 cm³ of aqueous sodium hydroxide and stir thoroughly. What is the colour of the precipitate obtained? Leave the mixture for at least half an hour and then observe what happens to the colour near the top of the mixture.

Can you explain what is happening? The experiment on the second portion may give you some ideas.

2. To the other portion, add an equal volume of chlorine water. Stir thoroughly and then warm the mixture. What happens to the colour of the solution? Add about 2 cm³ of aqueous sodium hydroxide. What colour is the precipitate which is obtained?

What are the names of the precipitates observed? Write ionic equations for the reactions observed. In what way is the chlorine acting in **2**?

What product would you expect if iron were to be reacted with chlorine?

Experiment 11.4 Reactions of copper compounds

You will need: copper(II) carbonate, dilute ammonia solution, aqueous sodium hydroxide, dilute nitric acid, dilute sulphuric acid. **Wear your safety glasses in this experiment.**

Divide the copper(II) carbonate into two portions.

1. a Heat one portion in a boiling tube and test for any gases produced. What happens to the colour of the solid?

b Allow the tube to cool and then add a few cm³ of dilute nitric acid. Warm the mixture gently until all the solid has dissolved. (You may need to add a little more acid.) What colour is the solution and what compound has been formed?

c Treat the solution with an equal volume of aqueous sodium hydroxide. What compound is formed in this reaction and what colour is it?

d Warm the mixture carefully and observe what happens. What compound is formed on heating?

Write chemical equations for the reactions observed in tests **a** to **d**.

2. a Place the other portion of copper(II) carbonate in a clean boiling tube and carefully add a few cm³ of dilute sulphuric acid. What gas is given off? Add some more acid if some solid does not dissolve. What is the name of the compound in solution? Write an equation for the reaction.

b To the solution add dilute aqueous ammonia. What compound is first formed? What happens when an excess of aqueous ammonia is added?

11.3 Metals and alloys—their uses

Zinc is widely used in dry cells and galvanising. Copper is an excellent conductor of electricity and is used for electrical wiring.

We have seen that metal elements have many useful properties, but pure metals are not widely used. The properties of metals such as strength and resistance to corrosion can be greatly improved by forming *alloys*.

An alloy is a mixture of one metal element with one or more other elements. The element mixed with the 'parent' metal is usually another metal or carbon.

How alloys are made

When alloys are made it is important that all the pure elements are completely mixed.

Some alloys are made directly by reducing mixed ores to form the alloy, but most alloys are made by mixing the pure elements in the correct proportions, when molten. The major ingredient is melted first and then the required quantities of the other elements are stirred in. The molten alloy formed is then poured into moulds to solidify.

By looking at a few common examples the advantages of forming alloys can be seen.

Brass is an alloy of copper with zinc.
Brass is stronger than copper and yet more easily worked.

Duralumin (mainly aluminium with magnesium and copper). This alloy is much stronger than aluminium. Its combination of high strength, low density, and resistance to corrosion, makes it ideal for pots and pans in kitchens and lightweight machinery. Aluminium alloys are also widely used in aircraft construction.

Steel is used to make car bodies and machinery.
Stainless steel is iron with chromium and nickel and a little carbon.
It is extremely hard wearing and resistant to corrosion, even when heated. Its resistance to corrosion can be improved by increasing the chromium content.

Questions

1. Which of the following reactions takes place in the manufacture of aluminum?

A Aluminium oxide is decomposed by heat.
B Aluminium oxide is reduced by heating it with carbon.
C Aluminium ions gain electrons to form aluminium atoms.
D The impurities in the aluminium ore are removed with limestone.
E Molten aluminium chloride is electrolysed.

2. Which one of the following metals reacts with steam, but not with water to give hydrogen?

A copper
B iron
C lead
D potassium
E sodium.

3. The following experiments were carried out on the elements P, Q, R and S:
   (i) S displaces P from an aqueous metal P salt
   (ii) P displaces Q from an aqueous metal Q salt
   (iii) Q displaces R from an aqueous metal R salt.

   What is the order of reactivity of these metals (the most reactive metal first)?

A P, Q, R, S
B R, P, Q, S
C S, P, Q, R
D Q, P, R, S
E P, Q, S, R.

4. A metal M was reacted with hydrochloric acid to give hydrogen. The hydrogen was passed over the heated oxide of a metal N. The oxide was reduced to the metal N.
   M and N could be:

A copper and lead
B lead and zinc
C zinc and copper
D copper and copper
E zinc and zinc.

5. When clean iron filings are placed in aqueous copper(II) sulphate:

A there is no reaction
B the copper(II) ions are oxidised
C iron(III) sulphate is precipitated
D iron atoms are oxidised
E sulphur is formed.

6. ALUMINIUM, BRASS, COPPER, IRON, POTASSIUM, MAGNESIUM, ZINC.

   Choose from the above list:

a the most reactive metal
b the least reactive metal
c a metal which bursts into flames when dropped into water
d a metal that does not react with dilute hydrochloric acid
e an alloy
f the metal extracted from the ore bauxite
g the metal used to galvanise iron
h the most plentiful metal in the earth's crust
i a metal which forms a brown oxide
j a metal which forms a blue sulphate.

7. Suggest reasons for the following:

a boats are never built of sodium
b saucepans are never made of lead
c mercury metal is used in thermometers
d tin cans are steel cans coated with tin
e lead is used as a weather seal on buildings
f aluminium is used for window frames
g copper is used for water pipes
h in high voltage electricity cables the current is carried by aluminium, but the cable has a core of steel

8. Decide if the following pairs of substances can react together. If they can, then write a word equation (or better still a balanced symbol equation) for the reaction:

a magnesium + oxygen
b copper + zinc nitrate solution
c iron(III) oxide + carbon monoxide
d lead(II) oxide + carbon
e magnesium oxide + zinc
f zinc + lead(II) nitrate solution
g copper(II) oxide and magnesium
h sodium + water
i copper + water
j silver + dilute hydrochloric acid
k magnesiun + dilute hydrochloric acid
l lead + silver nitrate solution
m magnesium + carbon dioxide

9. Use the following information to place the metals A, B, C and D in order of decreasing reactivity.

A does not react with dilute hydrochloric acid.
B forms a carbonate which does not decompose on heating.
C reacts with cold water, and forms a carbonate which loses carbon dioxide on heating.
D does not react with water, but does react when heated in steam.

10.

a What is an alloy?
b Why are alloys more widely used than pure metals?
c Name three alloys that can be found in your home. For each alloy:
 (i) name the elements it contains
 (ii) name the use to which it is put.

11. COPPER, SODIUM, ZINC, CALCIUM, LEAD.

a Arrange the above metals in order of decreasing reactivity.
b Which of the above metals can be extracted from their ores by electrolysis?
c Which of the above metals have oxides which can be reduced by hydrogen gas?
d Which of the above metals react with cold water?
e Which of the above metals forms a soluble carbonate?
f Which of the above metals forms a positive ion most easily?

12. Suppose you are given two unknown metals and solutions of their nitrates. Describe fully how you would find out which of the metals was the more reactive.

13. Place the metals calcium, copper and iron in order of decreasing activity (*most* reactive *first*) and justify your order by comparing

their reactions, if any, with:
 (i) water or steam
 (ii) dilute hydrochloric acid
Write equations for the reactions that you describe.

[C]

14. Aluminium is manufactured by the electrolysis of a molten mixture of aluminium oxide (Al_2O_3) and cryolite (Na_3AlF_6) in areas where electricity is cheap. The aluminium is formed at the cathode (negative electrode) and oxygen at the anode (positive electrode). The electrodes are made of graphite and the anode is gradually burnt away by the hot oxygen. Aluminium cannot be manufactured by the electrolysis of its chloride because aluminium chloride sublimes when heated; aluminium nitrate is also unsuitable because it decomposes into the oxide when heated.

a What is meant by *sublimes*? Name *two* other substances that sublime.
b Construct the equation for the action of heat on aluminium nitrate.
c Give reasons for the use of cryolite in the electrolysis of aluminium oxide.
d Suggest a reason for using graphite (rather than a metal) for the anode.
e Name *one* other metal that is manufactured by electrolysis. Name also *one* metal that is manufactured by reducing its oxide with carbon or carbon monoxide and write an equation for this reduction.
f Calculate the number of coulombs required to liberate:
 (i) 54 g of aluminium
 (ii) 54 g of silver.
Relative atomic masses: Al, 27; Ag, 108. The Faraday constant = 965000 C(coulombs)/mol.

[C]

15. Titanium(Ti) is the seventh most abundant metal in the Earth's crust. One of the most important ores of titanium is rutile(TiO_2).

 Titanium is manufactured from rutile in two stages as follows:

 (i) Chlorine is passed over a heated mixture of rutile and carbon giving titanium(IV) chloride ($TiCl_4$) and carbon monoxide. The titanium(IV) chloride (melting point $-23\,°C$; boiling point $136\,°C$) is condensed and purified by fractional distillation.
 (ii) Titanium(IV) chloride is reduced to titanium by heating it with a reactive metal such as magnesium in an atmosphere of argon. The solid titanium separates from the molten magnesium chloride (melting point $714\,°C$) which is tapped off.

 Titanium has a high mechanical strength and a low density and is very resistant to corrosion.

a Explain what is meant by (i) *an ore* (ii) *condensed* (iii) *fractional distillation*.

b In each case, suggest an element which could have been used in the manufacture of titanium (i) in place of magnesium (ii) in place of argon.

c In each case, construct the equation for (i) the preparation of titanium(IV) chloride (ii) the reduction of the chloride to the metal.

d Justify the use of the term *reduction* in **c** (ii).

e In each case, give the name and formula of any compound which has the same type of bonding as (i) titanium(IV) chloride (ii) magnesium chloride.

f Suggest a commercial use for titanium metal.

[C]

16.

a Outline the manufacture of iron from haematite (Fe_2O_3). Give equations for the reactions. (*No* diagram is required.)

b An *excess* of iron filings is added to an aqueous solution containing zinc sulphate and copper(II) sulphate. The mixture is shaken and filtered.
  (i) What ions are present in an aqueous solution containing zinc sulphate and copper(II) sulphate?
  (ii) What is the residue left after filtering?
  (iii) What ions are present in the filtrate?
  (iv) Write an ionic equation for any reaction taking place.

c Describe what you would have *observed* if you had carried out the experiment in **b**.

[C]

17. "Magnesium is a more reactive metal than iron."

a Describe carefully *two* reactions which you could carry out to confirm the statement above. Write equations for the reactions you describe.

b Name (i) *one* metal which is more reactive than *magnesium* (ii) *one* which is less reactive than *iron*.

c Magnesium chloride is an ionic solid. Give the formulae for the particles present in this solid. Under what conditions is magnesium chloride able to conduct an electric current?

d Calculate the maximum mass of iron that can be obtained from 80 tonnes of pure iron(III) oxide. Give one example of a commercial use of iron.

[C]

18. The following diagram shows a simplified sectional diagram of a blast furnace for the extraction of iron.

a (i) Name the substances fed into the furnace at A.

(ii) Name the substance(s) coming out of the furnace at B and at C.

b The waste gases contain carbon dioxide, carbon monoxide and nitrogen. Explain the presence of each of these gases.

c Iron exhibits allotropy. Explain what is meant by *allotropy* and name *one* other element that exhibits allotropy.

d Name a compound that is manufactured using iron as the catalyst. Write an equation for the catalysed reaction. Give *one* large scale use for this compound.

[C]

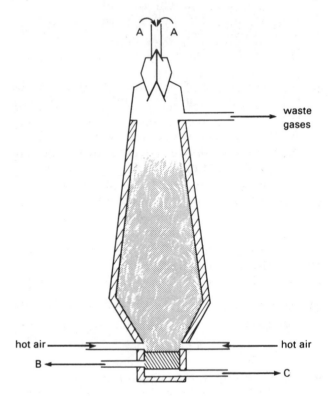

19. The following questions are all concerned with metals.

a Give the name of the metal which occurs most extensively in the earth's crust.

b Give the name of a metal which is extracted from its oxide by reduction with coke.

c (i) Give the name of a metal which is extracted by electrolysis.
  (ii) At which electrode are metallic ions discharged during electrolysis?

d (i) Give the name of a metal which will react vigorously with cold water.
  (ii) Give the name of a metal which will *not* react either with water or with dilute hydrochloric acid.
  (iii) Give the name of a metal which will evolve hydrogen when warmed *both* with hydrochloric acid *and* with aqueous sodium hydroxide.

e (i) Give the name or number of a Group in the Periodic Table which contains metals only.

(ii) Give the name or number of a Group in the Periodic Table which does not contain any metals.

(iii) Give the name of a metal which forms two series of compounds in which the metal shows different charges.

[L]

20. Excess iron filings were allowed to react with $50\,cm^3$ of $2\cdot0\,mol/dm^3$ hydrochloric acid according to the equation:

$$Fe(s) + 2HCl(aq) \rightarrow FeCl_2(aq) + H_2(g).$$

a (i) How would you show that the gas evolved was hydrogen?

(ii) Describe a test to show that iron(II) ions are present in the solution.

(iii) What is the colour of a concentrated solution of iron(II) chloride?

b (i) What volume of hydrogen gas, measured at room temperature and atmospheric pressure, would be evolved during the reaction?

(ii) What mass of hydrated iron(II) chloride ($FeCl_2.4H_2O$) would theoretically be produced in this experiment?

c (i) Write an ionic equation for the reaction between iron and hydrochloric acid.

(ii) How many moles of hydrogen ions are present in the $50\,cm^3$ of $2\cdot0\,mol/dm^3$ hydrochloric acid?

[L]

21. Metals are obtained from their compounds by reduction. Iron is produced using coke whereas the extraction of aluminium requires electrolysis.

a (i) Name *two* ores, one from which iron is obtained and the other from which aluminium is obtained.

(ii) How are the methods used for their extraction related to the positions of these two metals in the activity series?

b With reference to the extraction of aluminium,

(i) name the material used for the electrodes

(ii) write an equation for the formation of the product at the cathode

(iii) explain, with the aid of equation(s), why the anodes have to be replaced periodically.

c With reference to the extraction of iron,

(i) state the reactions which the coke undergoes

(ii) write an equation for the reaction by which the iron compound is converted to the metal

(iii) explain why limestone is added to the charge of iron ore and coke.

d Outline a process by which the iron obtained from the blast furnace is converted into steel.

[JMB]

22.

a Describe how you would prepare pure, dry samples of copper(II) oxide starting from:

(i) copper foil (ii) copper(II) carbonate.

Give an equation for one reaction involved in each case.

b Draw a labelled diagram of the apparatus which could be used to analyse the samples of the copper(II) oxide so prepared in order to determine the relative proportions by mass of copper and oxygen in them.

State any precautions you would take to ensure safety and an accurate result in this determination.

[JMB]

12 Hydrogen

An atom of the common isotope of hydrogen contains only 1 proton, 1 electron and no neutrons. This means hydrogen is the element with the lightest atoms. Hydrogen gas exists as H_2 molecules by sharing electrons. It is the least dense of all gases.

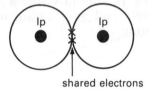

shared electrons

Fig 12.1 A hydrogen molecule

12.1 The uses of hydrogen

Hydrogen is a very important element. It has a large number of important industrial uses.

1. Hydrogen is used to manufacture ammonia gas

nitrogen + hydrogen $\rightleftharpoons$ ammonia
$$N_2 \;+\; 3H_2 \;\rightleftharpoons\; 2NH_3$$

This is the major use of hydrogen, as very large quantities of ammonia are needed to make fertilisers, dyes and plastics.

2. Hydrogen is used in the manufacture of margarine. It is used to change vegetable oils, such as palm oil or olive oil into solid fats.

3. Hydrogen is used to provide a reducing atmosphere in furnaces.

Most metals, when hot, react with oxygen in the air. They become tarnished with a layer of metal oxide. By having an atmosphere of hydrogen in the furnace, tarnishing can be prevented. In this way metals can be annealed or brazed more efficiently.

12.2 Making hydrogen in schools

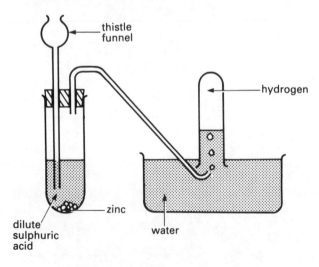

Fig 12.2 Making hydrogen in the laboratory

When we want to make small amounts of hydrogen gas in school laboratories, most of the industrial methods are not suitable. Hydrogen is usually made in laboratories by reacting zinc with dilute hydrochloric or sulphuric acid.

zinc + sulphuric acid $\rightarrow$ zinc sulphate + hydrogen
$$Zn(s) + \; H_2SO_4(aq) \; \rightarrow \; ZnSO_4(aq) \; + \; H_2(g)$$

Pure zinc reacts very slowly with dilute sulphuric acid. By adding a little copper(II) sulphate solution to the reaction mixture, hydrogen is produced much faster.

Hydrogen is only slightly soluble in water and so it is usually collected over water. Since it is far less dense than air it can be collected by upward delivery as shown in Fig. 12.3.

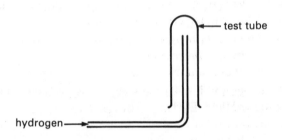

Fig 12.3 Collecting hydrogen by upward delivery

If a dry sample of hydrogen is needed it must be collected by upward delivery. Hydrogen gas can be dried by passing it through any drying agent, such as concentrated sulphuric acid, silica gel or anhydrous calcium chloride. The problem when collecting hydrogen by upward delivery is that you cannot tell when the test tube is full of hydrogen.

12.3 Hydrogen—the chemical

Hydrogen is a colourless, tasteless gas with no smell. It is the first element in the Periodic Table. Hydrogen atoms are far too unstable to exist by themselves, so hydrogen is found as H_2 molecules.

Chemical properties of hydrogen

1. Mixtures of hydrogen and oxygen burn very readily to form water:

hydrogen + oxygen $\rightarrow$ water
$$2H_2(g) \; + \; O_2(g) \rightarrow 2H_2O(g)$$

The reaction can often be explosive, so great care has to be taken when burning hydrogen gas. We use this burning reaction as a test for hydrogen.

Test for hydrogen: *mixtures of hydrogen and air burn, usually with a squeaky pop.*

Because of the risk of an explosion, you should

never try lighting more than a test tube full of hydrogen.

2. Hydrogen will react with chlorine to form hydrogen chloride.

hydrogen + chlorine → hydrogen chloride
$$H_2(g) + Cl_2(g) → 2HCl(g)$$

This reaction is best seen by lowering a jet of burning hydrogen into a gas jar of chlorine.

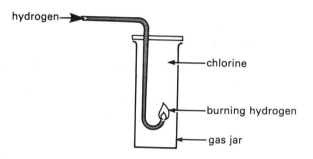

Fig 12.4 Burning hydrogen in chlorine

When burning hydrogen is lowered into a gas jar of chlorine the hydrogen continues to burn with a white flame. The green colour of the chlorine slowly disappears. When all the chlorine has reacted the hydrogen stops burning.

Note: The reaction between hydrogen and chlorine can sometimes be violently explosive. It should never be attempted by a pupil.

3. Hydrogen will take the oxygen from the oxides of some metals. The more unreactive metals such as iron, lead and copper can be obtained in this way. Hydrogen is acting as a reducing agent in these reactions.

Example:

copper(II) oxide + hydrogen → copper + water
$$CuO + H_2 → Cu + H_2O$$

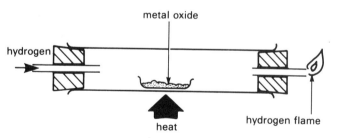

Fig 12.5 Reacting hydrogen with metal oxides

4. Pure hydrogen is a neutral gas. It has no effect on litmus paper or Universal indicator paper.

Questions

1. Which one of the following gases diffuses most rapidly?

A H_2
B HCl
C H_2S
D NH_3
E CH_4

2. Hydrogen burns in chlorine to form hydrogen chloride. Which one of the following statements about this reaction is not true?

A Hydrogen chloride has been synthesised.
B The hydrogen chloride formed burns in air.
C The reactants and products are all gases.
D A colour change is seen.
E The reaction is exothermic.

3. If $10\,cm^3$ of hydrogen was burned in $10\,cm^3$ of oxygen at $100\,°C$ and atmospheric pressure, the product would be:

A $5\,cm^3$ of hydrogen and $10\,cm^3$ of steam
B $10\,cm^3$ of hydrogen and $10\,cm^3$ of steam
C $10\,cm^3$ of steam only
D $5\,cm^3$ of oxygen and $10\,cm^3$ of steam
E $10\,cm^3$ of oxygen and $10\,cm^3$ of steam.

4. Which one of the following metal oxides can be reduced by hydrogen?

A aluminium oxide
B calcium oxide
C lead(II) oxide
D lithium oxide
E magnesium oxide

5. Which one of the following is the ionic equation for the reaction between zinc and dilute sulphuric acid?

A $Zn(s) + H^+(aq) → Zn^+(aq) + H(g)$
B $Zn(s) + 2H^+(aq) → Zn^{2+}(aq) + H_2(g)$
C $Zn(s) + H_2^+(aq) → Zn^+(aq) + H_2(g)$
D $Zn(s) + 2H^+(aq) → Zn^{2+}(aq) + 2H(g)$
E $Zn(s) + 2H_2^+(aq) → Zn^{2+}(aq) + 2H_2(g)$

6.

a When copper is heated in air, it tarnishes, and copper(II) oxide is formed. Write an equation for the reaction.

b When copper(II) oxide is heated in hydrogen, copper and water are formed. Write an equation for the reaction.

c When copper or brass components are brazed together in a furnace, an atmosphere of hydrogen is often kept in the furnace. Why is the hydrogen used?

d What safety rules would be needed in a factory using hydrogen in this way?

7. Suggest reasons for the following:
a A balloon filled with hydrogen rises.
b A hydrogen filled balloon deflates faster than an air filled balloon.
c Hydrogen can be made by reacting zinc with dilute sulphuric acid but this method is not used to make hydrogen industrially.

8. A student wanted to prepare a few test tubes of hydrogen gas. He drew the following diagram of the apparatus he intended to use:

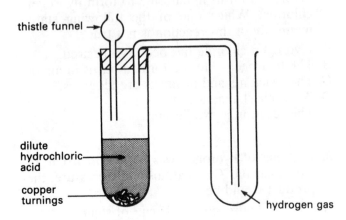

thistle funnel →

dilute hydrochloric acid

copper turnings

hydrogen gas

a Name three mistakes that the student made in his diagram.
b How would you prove that a test tube contained hydrogen gas?

9. Water and hydrogen chloride (HCl) are covalent substances. Draw diagrams to show how the electrons are arranged in molecules of these compounds.

10.
a When hydrogen burns in air, water is formed. Write an equation for the reaction.
b Draw a diagram of the apparatus you would use to collect some of the water formed when hydrogen burns. (You need not show how the hydrogen was prepared.)
c How would you show that the liquid you collect is water?

11.
a Give a diagram to show how a sample of hydrogen may be prepared and collected in the laboratory. Write the equation for the reaction involved.
b Explain why the method you describe in a is not suitable for industrial use.
c State the essential reaction conditions and write the equations for the reactions of hydrogen with: chlorine, ethene, nitrogen.
d When 5·40 g of an oxide of uranium (U) were reduced, 4·76 g of uranium metal were obtained. Calculate the simplest formula for the oxide.
(Relative atomic masses: O, 16·0; U, 238)
[C]

12.
a Give the reagents and conditions necessary for the preparation of a sample of hydrogen in the laboratory by a process involving: electrolysis, or a method other than electrolysis.
In each case, write the equation for the reaction in which hydrogen is produced.
b (i) 6·0 g of an oxide of lead gave 5·2 g of lead, Pb, on reduction by hydrogen. Calculate the empirical formula of the oxide. (Relative atomic masses: O = 16, Pb = 208.)
(ii) Give the name of another oxide which can be reduced to the metal by hydrogen.
(iii) Give the name of an oxide of a metal which cannot be reduced to the metal by hydrogen.
c (i) How, and under what conditions, does hydrogen react with ethene?
(ii) In the margarine industry, use is made of this type of reaction to turn natural oils into fats. From this statement, draw a conclusion about the bonding in natural oil molecules.
d As there are vast, untapped reserves of hydrogen in the waters of the world and its use as a fuel carries no pollution risk, it has been said that hydrogen is the fuel of the future. Give one reason in each case why:
(i) we have not obtained hydrogen on a large scale from this source,
(ii) its use as a fuel carries no pollution risk.
[C]

13.
a By means of a labelled diagram and an equation for the reaction involved, show how you would prepare and collect gas jars of hydrogen.
b Describe fully how you would prepare, by direct synthesis, from hydrogen a sample of an acidic gas. Give the name and formula of this compound.

14. Describe the preparation and collection of samples of hydrogen from water. Draw a diagram of the apparatus you would use.

15. The progress of the reaction
$$Zn + H_2SO_4 \rightarrow ZnSO_4 + H_2$$
may be followed by recording the total volume of hydrogen evolved after known times.
a Give a diagram of an apparatus suitable for carrying out the experiment. Sketch a graph showing the type of results you would expect to obtain. Label the axes of your graph as 'total volume of hydrogen' (vertical axis) and 'time from start of reaction' (horizontal axis).

b In a second experiment, a more concentrated solution of sulphuric acid was used, and the reaction was found to proceed faster. What explanation can you give for this observation? Name *two* other changes in the experimental conditions that you would also expect to lead to a faster reaction.

c What would be the maximum volume of hydrogen evolved if 6·5 g of zinc and 100 cm³ of 0·10 mol/dm³ sulphuric acid were used?

(Relative atomic mass: Zn, 65. Assume that one mole of hydrogen occupies 24 dm³ under the conditions of the experiment.)

[C]

13 The air and oxygen

13.1 Air: What is in it?

Air is a mixture. It is a most important mixture. Without it, life as we know it could not exist. Table 1 shows what normal air contains.

Name of substance	Percentage in air (by volume)
nitrogen	78%
oxygen	20%
carbon dioxide	0·03%
noble gases	1% (mainly argon)
water vapour	variable (often about 1%)
polluting gases	variable

Table 1

Because air is a mixture, the percentage of each gas varies from time to time and from place to place.

Nitrogen

The largest part of air is nitrogen. It is an unreactive element. It can react with oxygen during thunderstorms to form nitrogen dioxide and this can dissolve in rain water to form nitric acid.
Most of the time the nitrogen in the air shows no sign of reacting.

Oxygen

Oxygen is the most active part of the air. We need oxygen for breathing and for burning all fuels.

Carbon dioxide

Only a small fraction of the air is carbon dioxide, but it is a very important part. It is needed by plants to make food.

Noble gases

These are helium, neon, argon, xenon, krypton and radon. They are very unreactive but they are very useful to man once they have been separated from the air.
Helium is used in helium/oxygen mixtures for diving. It is much better than compressed air for deep diving.

Neon is used to fill some lamp bulbs and advertising strip lights.
Argon is the most plentiful of the noble gases in air. It is used to fill light bulbs. Do you know why light bulbs are not filled with air?

Water vapour

The water vapour in the air varies from time to time. Imagine the air in your kitchen. It contains far more water vapour when food is being prepared that at other times.

Polluting gases

These are the harmful gases in the atmosphere. They get there mainly by man's activity. Table 2 shows some of the common polluting gases which are found in air.

13.2 Getting chemicals from air

Air is a convenient source of important gases needed by industry. These gases are obtained by the fractional distillation of liquid air. There are a number of stages in this process.

1. The air is filtered to remove dust.
2. Carbon dioxide is removed by passing the air through sodium hydroxide solution.
3. Water vapour is removed using a drying agent.
4. The air is compressed to about 200 atmospheres and then cooled.
5. The cooled compressed air is then allowed to expand suddenly. This sudden expansion cools the air even more and liquid air is eventually formed at about $-200\,°C$.
6. Finally, the liquid air is fractionally distilled. Pure nitrogen, oxygen and argon can be obtained.

Once these gases have been separated from air, they are sold either as gases in cylinders, or as liquids. Enormous quantities of nitrogen and oxygen are used by industry each year.

13.3 Oxygen—the V.I.P.

Oxygen is a very important element. We need it for breathing. When we breathe we draw air into

Name of polluting gas	How it gets into the atmosphere	Harmful properties
sulphur dioxide	by burning fuels.	poisonous to man. acidic, attacks buildings and railings.
nitrogen dioxide	from car exhausts. produced by some industries	poisonous to man. acidic.
carbon monoxide	from car exhausts	poisonous to man.

Table 2

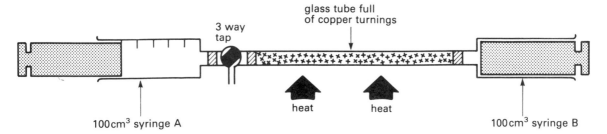

Fig 13.1 Finding out the percentage of oxygen in air

our lungs. In our lungs red blood cells pick up oxygen. They can do this because they contain the substance *haemoglobin*. Haemoglobin reacts with oxygen to form a substance called *oxy-haemoglobin*.

HAEMOGLOBIN + OXYGEN
$\rightleftharpoons$ OXYHAEMOGLOBIN

The red blood cells then travel around the body. At various parts of the body oxyhaemoglobin gives up its oxygen where it is required.

Our lungs are not very efficient. They only use about $\frac{1}{4}$ of the oxygen we breathe in. When people are ill sometimes their lungs work even less efficiently. We give these people extra oxygen from cylinders, so that their bodies can more easily get the oxygen they need.

The pollutant carbon monoxide is poisonous because it readily combines with the haemoglobin in the lungs. When this happens, the oxygen cannot be absorbed and transported around the body.

Fish also need oxygen. They obtain it from the air that is dissolved in the water. Oxygen is more soluble than nitrogen in water and dissolved air contains about 30% oxygen.

You can find out the percentage of oxygen in the air in your school laboratory by using the apparatus shown in Fig 13.1.

Draw 100 cm³ of air into syringe A through the three way tap. Turn the three way tap so that the air in syringe A can be passed into syringe B. Heat the copper strongly and pass the air backwards and forwards between syringe A and syringe B. The hot copper reacts with the oxygen in the air to form black copper(II) oxide.

copper + oxygen $\rightarrow$ copper(II) oxide
2Cu + O$_2$ $\rightarrow$ 2CuO

When no more copper(II) oxide is being formed stop heating. Push the remaining air back into syringe A and let the apparatus cool. If you try this experiment you will most likely find that you are left with about 80 cm³ of air in syringe A.

So, 100 cm³ of air contains $100 - 80 = 20$ cm³ of oxygen.

The percentage of oxygen in the air
$$= \frac{20}{100} \times 100 = 20\%$$

How would you change this experiment to show that the air you breathe out contains only 16% oxygen?

13.4 Making oxygen

We have seen that oxygen can be separated from the air. It can be bought compressed in cylinders. However, in schools, we often need a very small amount of oxygen for experiments. It is simpler to make some oxygen when it is needed rather than rent or buy a large cylinder.

By breaking down hydrogen peroxide

Hydrogen peroxide is a fairly unstable chemical. It easily breaks down to form water and oxygen:

hydrogen peroxide $\rightarrow$ water + oxygen
2H$_2$O$_2$(aq) $\rightarrow$ 2H$_2$O(l) + O$_2$(g)

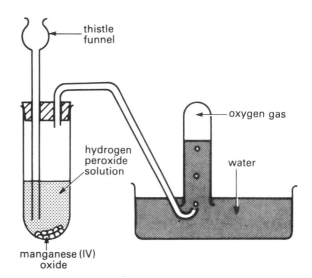

Fig 13.2 Making oxygen

The manganese(IV) oxide makes the hydrogen peroxide break down quickly at room temperature. The manganese(IV) oxide is acting as a catalyst.

Oxygen from plants

The green colouring in plants is due to a substance called *chlorophyll*. This is the catalyst for a reaction between carbon dioxide (from the air) and water (absorbed through the roots).

Light energy is needed for the reaction to proceed. The products are glucose and oxygen. This process is called *photosynthesis*.

$$\text{carbon dioxide} + \text{water} \xrightarrow[\text{chlorophyll}]{\text{light}} \text{glucose} + \text{oxygen}$$

$$6CO_2 + 6H_2O \longrightarrow C_6H_{12}O_6 + 6O_2$$

This sugar (glucose), can then be converted into starch and other carbohydrates.

In the generation of electricity, most power stations burn enormous amounts of coal, oil or gas. This process uses vast quantities of oxygen from the air and produces vast quantities of carbon dioxide. You can understand why large areas of forest are important if the air is to maintain a balance between oxygen and carbon dioxide.

13.5 Oxygen—the chemical

Oxygen exists as O_2 molecules. It is a colourless gas with no smell. It is not very soluble in water, so it is usually collected over water.

Oxygen reacts with most other elements to form oxides. One method of reacting solid elements with oxygen is shown in Fig 13.3. A sample of the solid element is heated on a combustion spoon and the combustion spoon is then lowered into a gas jar of oxygen.

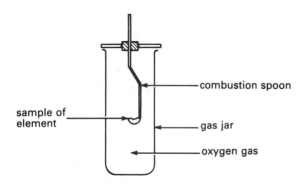

Fig 13.3 Burning elements in oxygen

Table 3 shows what happens when a number of common elements react with oxygen in this way.

The elements in Table 3 also react with oxygen when heated in air, but the reactions are not as violent. When oxygen reacts with elements in this way, heat is always produced. The reactions are always *exothermic*.

The oxides of metals have different properties from the oxides of non-metals.

Metal oxides

These are high melting point solids. They neutralise acids and are therefore known as *basic oxides*. Most do not dissolve in water. Metal oxides that dissolve in water form alkaline solutions.

Aluminium oxide, lead(II) oxide and zinc oxide can also react with alkalis. They are therefore called *amphoteric oxides* (see Chapter 9).

Non-metal oxides

These usually have low melting points. Many are gases. Most are soluble in water. They dissolve to form acidic solutions. These oxides are known as *acidic oxides*. There are a few non-metal oxides that do not have acidic properties. These are known as *neutral oxides*. Nitrogen monoxide (NO) and carbon monoxide (CO) are examples of neutral oxides.

Corrosion of metals

Metals corrode because they react with oxygen to form oxides. The most common example is the rusting of iron. However, the rusting process is *not* just a reaction between iron and oxygen as the following experiments show.

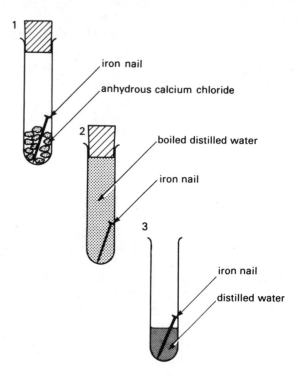

Fig 13.4

In tube 1, the iron nail is in contact with air (containing oxygen) but no water is present. Rusting does not occur.

In tube 2 the nail is in contact with water but not with oxygen. Rusting does not occur.

In tube 3 the nail is in contact with both oxygen and water. Rusting occurs readily.

From these simple experiments we can see that *both* oxygen and water must be present for rusting to occur.

The prevention of rusting

This would seem to be fairly easy—stop oxygen and water from reaching the surface of the iron.

Most objects made of iron or steel are usually coated to prevent reactions occurring at the metal surface. Paint or plastic are often used. Unfortunately, most coatings tend to peel off after a time.

NAME OF ELEMENT	EXPERIMENT	REACTION
Phosphorus	If a small piece of white phosphorus is warmed very gently and then lowered into oxygen it bursts into flames. It burns with a white flame and makes a thick white smoke. After some time a fine white powder settles in the gas jar.	phosphorus + oxygen → phosphorus(V) oxide $2P$ + $5O_2$ → P_2O_5
Sulphur	If a sample of sulphur is heated until it just melts and is then lowered into oxygen it bursts into flames. It burns with a bright blue flame. A colourless gas with a choking smell is formed.	sulphur + oxygen → sulphur dioxide S + O_2 → SO_2
Carbon	If a sample of carbon is heated strongly (until just red hot) and then lowered into oxygen it *glows far more brightly.* A colourless gas is formed.	carbon + oxygen → carbon dioxide C + O_2 → CO_2
Magnesium	If a strip of magnesium ribbon is heated until it just starts to burn and is then lowered into oxygen it burns very strongly. A blinding white light is produced. A white powder is formed.	magnesium + oxygen → magnesium oxide $2Mg$ + O_2 → $2MgO$
Iron	If some iron wool is heated until red hot and then lowered into oxygen it glows white hot. Sparks fly in all directions. A black powder is formed.	iron + oxygen → tri-iron tetraoxide $3Fe$ + $2O_2$ → Fe_3O_4

Table 3

Another method is to attach a more reactive metal to the iron surface. The hulls of ships often have blocks of magnesium attached to them. As it is more reactive than iron, the magnesium corrodes rather than the hull of the ship.

In a similar way zinc is often used to coat steel or iron objects. This process is known as *galvanising*. The layer of zinc prevents oxygen and water from reaching the iron surface. Should the layer of zinc get scratched, then the zinc will corrode because zinc is the more reactive metal.

Test for oxygen

If a glowing splint is placed in oxygen it will relight. This is the usual test for oxygen.

Uses of oxygen

The main use is in steel making (Chapter 11). Large quantities of oxygen are needed. Many large steelworks have plants for making oxygen from air. Oxygen is used with ethyne (acetylene) for welding, brazing and metal cutting. The use of oxygen to help patients' breathing in hospitals has already been mentioned.

Questions

1. The boiling points of various gases in the air are:

argon $-186\,°C$, krypton $-153\,°C$, neon $-246\,°C$, nitrogen $-196\,°C$, oxygen $-183\,°C$.

If air is cooled, the first substance obtained is water followed by carbon dioxide. If the temperature is lowered further, the next substance obtained is:

A argon   D nitrogen
B krypton   E oxygen.
C neon

2. Which one of the following is the best substance to use to remove water vapour from moist oxygen?

A aluminium oxide
B anhydrous copper(II) sulphate
C concentrated sulphuric acid
D lime water
E sodium hydroxide solution

3. Oxygen was prepared and collected as shown.

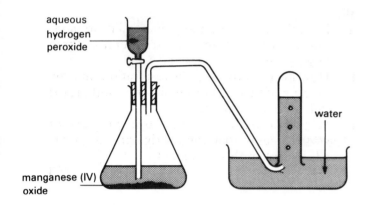

aqueous hydrogen peroxide

manganese (IV) oxide

water

The first few test-tubes of gas were rejected because the oxygen also contained:

A air
B hydrogen
C hydrogen peroxide
D manganese(IV) oxide
E water vapour.

4. For iron to rust, it must be in contact with:

A oxygen and water
B nitrogen and water
C oxygen and nitrogen
D oxygen and carbon dioxide
E nitrogen and carbon dioxide.

5.
a Name three gases present in normal air.
b Name three gases that pollute air.
c Give three reasons why air is a mixture and not a compound.

6. ARGON, CARBON DIOXIDE, CARBON MONOXIDE, HELIUM, OXYGEN, NITROGEN.
Choose from the above list:

a A gas used to make steel from molten iron.
b A gas used to fill light bulbs.
c A gas formed when wood burns.
d A gas which is more dense than air.
e Two gases present in car exhaust fumes.
f The most plentiful gas in the Earth's atmosphere.
g A gas used up during breathing.
h A gas formed during breathing.

7. Draw a fully labelled diagram of the apparatus you would use to prepare a few test tubes of oxygen gas starting from hydrogen peroxide solution.

a How would you test the gas you collected to show it was oxygen?
b Write an equation for the reaction involved in this preparation.
c What mass of hydrogen peroxide would you need to use to prepare 1 mole of oxygen gas?

8. The apparatus below was designed to demonstrate that magnesium reacts with nitrogen from the air to form magnesium nitride, Mg_3N_2.

Tubes 1 and 2 contained fine copper wire and tube 3 contained fine magnesium powder. A slow stream of dry air from which all the carbon dioxide had been removed was passed through the apparatus and tubes 1, 2 and 3 were heated.

In tube 1, the copper wire turned black.

In tube 2, the copper wire remained unchanged.

In tube 3, the magnesium powder turned pale yellow.

In beaker 4, the bubbles decreased to a small rate of flow.

When water was added to the cold product from tube 3, magnesium hydroxide was formed and a pungent gas was given off which turned damp litmus paper blue.

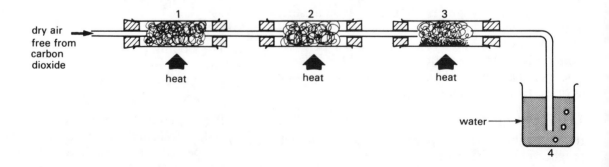

dry air free from carbon dioxide

heat   heat   heat

water

a Draw a diagram to show how dry air free from carbon dioxide could be obtained.
b What is the chemical reaction in tube 1?
c What is the purpose of tube 2?
d Name the gases given off during, and write equations for, the action of water on: (i) magnesium powder, (ii) magnesium nitride.
e Identify the bubbles of gas in beaker 4.
f Explain why you would expect the rate of bubbling in beaker 4 to increase rapidly after a period of time.

[C]

9.
a (i) Name *three* substances, other than oxygen and nitrogen, that are always present in the atmosphere.
  (ii) Describe, with the aid of a diagram of the apparatus, an experiment by which you could demonstrate the presence in the atmosphere of *one* of these three substances.

b Describe, with the aid of a diagram of the apparatus, the preparation and collection of a sample of oxygen, starting with hydrogen peroxide.
c Outline briefly how air is liquefied. How is oxygen obtained from liquid air?
d State the approximate percentage of oxygen by volume in the atmosphere, and explain briefly why the figure does not vary very much.

[C]

10.
a Describe in outline the commercial preparation of oxygen from the air. Mention three large-scale uses of oxygen.
b How do the processes of respiration and the burning of fuels: (i) resemble one another, (ii) differ from one another?
c 'When we burn coal, we are making use of stored energy that came originally from the sun.' Explain this statement.

[C]

14 Nitrogen and its compounds

There is a lot of nitrogen about. 78% of the Earth's atmosphere is nitrogen. It exists as N_2 molecules. The two atoms in the molecule are very tightly held together. Because of this, nitrogen does not take part in many chemical reactions. We say nitrogen is an *inert* substance. Many of the uses of nitrogen gas depend on it being inert. For example, nitrogen is used with argon to fill electric light bulbs. These gases are used because they do not react with the tungsten filament, even when it is white hot.

14.1 Nitrogen in compounds

Although nitrogen gas does not take part in many chemical reactions, a large number of very important compounds contain nitrogen. It is present in all protein. All plants and animals contain protein.

Animals get their protein by eating plants or other animals but plants make their own protein. For plants to make protein they need a supply of nitrogen.

Although plants do get their nitrogen from the soil, they cannot absorb nitrogen gas. They absorb water soluble compounds, which contain nitrogen. These compounds, usually nitrates, are absorbed through the roots.

As plants remove nitrates from the soil to make protein, the soil becomes less and less fertile. The nitrates have to be replaced.

Farmers use nitrogen fertilisers to keep a high level of nitrates in the soil. The most common nitrogen fertilisers are ammonium sulphate and ammonium nitrate.

14.2 Making nitrogen gas more useful

There are natural deposits of substances such as sodium nitrate and potassium nitrate. These have been widely used as fertilisers in the past, but as farming became more intensive the demand for these substances outstripped the supply. There was a need to convert nitrogen gas into nitrogen fertilisers. Most nitrogen fertilisers are nowadays made from ammonia. The Haber process made it possible to manufacture ammonia from nitrogen gas.

The Haber process for manufacturing ammonia

In the Haber process nitrogen is made to react with hydrogen to form ammonia gas:

nitrogen + hydrogen $\rightleftharpoons$ ammonia
$$N_2 \quad + \quad 3H_2 \quad \rightleftharpoons \quad 2NH_3$$

Nitrogen is obtained from the air and hydrogen is obtained from natural gas or water.

To get the nitrogen and hydrogen to react together, the gases have to be compressed to 150–200 atmospheres pressure and passed over an *iron* catalyst at 380–450 °C. Under these conditions about 25% of the nitrogen and hydrogen is changed into ammonia. This ammonia is separated from the nitrogen and hydrogen by liquefying the ammonia. The nitrogen and hydrogen are passed over the catalyst again.

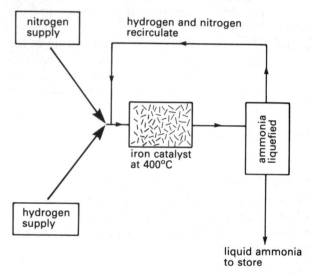

Fig 14.1 The Haber process

You will have seen from the equation that the reaction between nitrogen and hydrogen is reversible. At the same time as nitrogen and hydrogen are reacting to form ammonia, the ammonia is breaking up to reform nitrogen and hydrogen. A high pressure is used in the Haber process as this gives a greater amount of ammonia in the mixture leaving the catalyst chamber.

Uses of ammonia

The ammonia produced in the Haber process is sold either as a compressed gas or as a liquid. Most of this ammonia is used to make nitrogen fertilisers such as ammonium sulphate and ammonium nitrate.

The other main uses of ammonia gas are to manufacture nitric acid and nylon.

You should realise from its uses that ammonia is a very important chemical.

14.3 Ammonia—the chemical

We could not hope to make ammonia in the laboratory by a Haber process method. Instead it is prepared by heating an ammonium salt with an alkali. Ammonium chloride and calcium hydroxide are often used:

calcium	+	ammonium	→	calcium	+	ammonia	+	water
hydroxide		chloride		chloride				

$$Ca(OH)_2 + 2NH_4Cl \rightarrow CaCl_2 + 2NH_3 + 2H_2O$$

Flask A is sloped as shown in Fig 14.2 so that any water formed will run away from the hot flask.

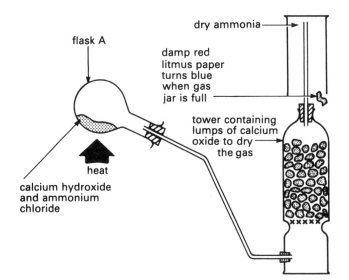

Fig 14.2 Making ammonia in the laboratory

Physical properties of ammonia

Ammonia is a colourless gas with a strong smell. It is less dense than air and is *very* soluble in water. Because of this ammonia has to be collected by upward delivery.

Chemical properties of ammonia

Ammonia is a very reactive gas. It takes part in a large number of chemical reactions.

1. *With water*

Ammonia is very soluble in water. Great care has to be taken when dissolving the gas in water.

If ammonia is to be safely dissolved in water some method must be used to prevent the water sucking back into the reaction flask. Fig 14.3 shows a simple method used to dissolve ammonia in water safely in the laboratory.

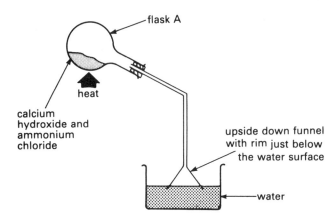

Fig 14.3 A safe method for dissolving ammonia in water

The funnel must be arranged with its rim just below the water surface. As the ammonia dissolves, the water starts to suck back in the funnel. As it does so the water level in the trough goes down. This leaves a plug of water in the wide part of the funnel. This plug cannot be supported and so it falls down. In this way suck back is prevented.

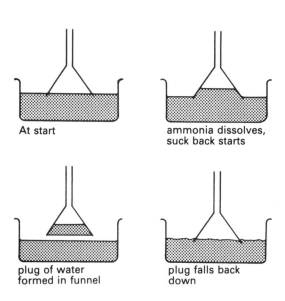

Fig 14.4

When ammonia dissolves in water it reacts with the water to form ammonia solution which is a weak alkali:

ammonia	+	water	⇌	ammonia solution	⇌	ammonium ions	+	hydroxide ions

$$NH_3(g) + H_2O(l) \rightleftharpoons NH_3(aq) \rightleftharpoons NH_4^+(aq) + OH^-(aq)$$

2. *Ammonia as an alkali*

Ammonia is an alkaline gas. It will turn damp red litmus paper blue. *This can be used as a test for the gas.*

Because ammonia is an alkaline gas it will react with acids to form salts. This is the reason why ammonia cannot be dried using concentrated sulphuric acid. The acid would react violently with the ammonia:

ammonia	+	sulphuric acid	→	ammonium sulphate

$$2NH_3 + H_2SO_4 \rightarrow (NH_4)_2SO_4$$

When ammonia reacts with acids, ammonium salts are formed. One of these reactions is also used as a test for ammonia gas. If ammonia gas is brought into contact with hydrogen chloride gas, a thick white smoke of ammonium chloride is formed. Concentrated hydrochloric acid can be used as a source of hydrogen chloride:

ammonia	+	hydrogen chloride	→	ammonium chloride

$$NH_3(g) + HCl(g) \rightarrow NH_4Cl(s)$$

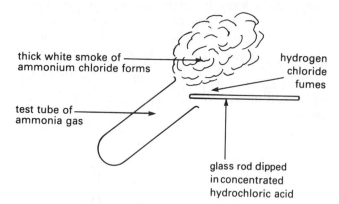

thick white smoke of ammonium chloride forms

hydrogen chloride fumes

test tube of ammonia gas

glass rod dipped in concentrated hydrochloric acid

Fig 14.5 Testing for ammonia gas

Experiment 14.1 Comparing the properties of aqueous ammonia with aqueous sodium hydroxide

You will need: aqueous solutions of ammonia, sodium hydroxide, aluminium sulphate, calcium nitrate, copper(II) sulphate, iron(II) sulphate, iron(III) chloride, lead(II) nitrate and zinc sulphate.

Place about 2 cm depth of aqueous aluminium sulphate in a test tube. Add aqueous ammonia until it is present in excess. Repeat the experiment using sodium hydroxide in the place of aqueous ammonia.

Note the colour of the precipitates formed, and see if the precipitate is soluble when excess alkali is added.

Repeat the experiments using the other salts provided.

You should now be able to identify certain metal ions by testing them with aqueous ammonia and aqueous sodium hydroxide. (See Chapter 19.)

Solutions of ammonia in water, being alkaline, behave like solutions of sodium hydroxide in water. They can be used to precipitate metal hydroxides from solutions of metal salts.

Example:

iron(II) sulphate	+	ammonia solution	→	iron(II) hydroxide	+	ammonium sulphate

$$FeSO_4(aq) + 2NH_4OH(aq) \rightarrow Fe(OH)_2(s) + (NH_4)_2SO_4(aq)$$
dirty blue-green precipitate

Because solutions of ammonia in water also contain some free ammonia molecules, some differences are noticed. Both ammonia solution and sodium hydroxide solution produce a pale blue precipitate with copper(II) sulphate solution. However, this precipitate dissolves in excess ammonia solution, to form a dark blue solution of a complex salt. This does not happen with sodium hydroxide solution. Zinc hydroxide will also dissolve in an excess of aqueous ammonia because of the formation of a complex salt.

3. Burning ammonia

Ammonia will not burn in air, but it will burn in oxygen to form nitrogen and water.

ammonia + oxygen → nitrogen + water
$$4NH_3 + 3O_2 \rightarrow 2N_2 + 6H_2O$$

This is not an important reaction because the ammonia burnt is of far greater use than the nitrogen formed. However, when a platinum catalyst is present, ammonia and oxygen can be made to undergo a far more important and useful reaction.

ammonia + oxygen $\xrightarrow[\text{and heat}]{\text{platinum catalyst}}$ nitrogen monoxide + water

$$4NH_3 + 5O_2 \longrightarrow 4NO + 6H_2O$$

This reaction plays a very important part in the manufacture of nitric acid.

4. Ammonia as a reducing agent

If ammonia gas is passed over heated copper(II) oxide a reaction takes place. The copper(II) oxide is reduced to copper by losing oxygen. Ammonia is changed into nitrogen and water. It is oxidised:

ammonia + copper(II) oxide → copper + water + nitrogen

$$\underset{\text{oxidised}}{2NH_3(g)} + \underset{\text{reduced}}{3CuO(s)} \rightarrow 3Cu(s) + 3H_2O(l) + N_2(g)$$

The ammonia is the reducing agent.

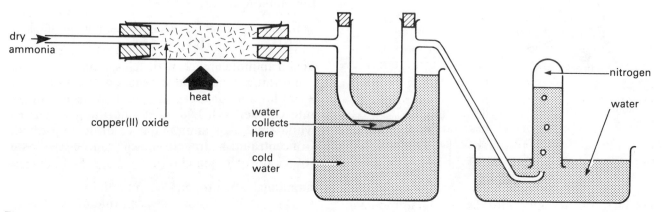

dry ammonia

heat

copper(II) oxide

water collects here

cold water

nitrogen

water

Fig 14.6 Reacting ammonia and copper(II) oxide

14.4 Ammonium salts

Ammonium salts are the compounds formed when acids are neutralised by ammonia. They are ionic solids which are soluble in water. Some, like ammonium chloride, break down on heating and reform on cooling. They *thermally dissociate*.

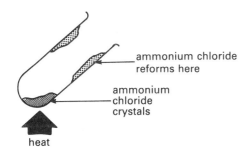

Fig 14.7 Thermal dissociation of ammonium chloride

When ammonium chloride is heated it does not melt, but changes directly into a colourless gas. It appears to sublime. Ammonium chloride is formed on the cooler parts of the test tube.

ammonium $\underset{\text{cool}}{\overset{\text{heat}}{\rightleftharpoons}}$ ammonia + hydrogen
chloride   chloride

$$NH_4Cl(s) \underset{\text{cool}}{\overset{\text{heat}}{\rightleftharpoons}} NH_3(g) + HCl(g)$$

Many ammonium salts are widely used. Table 1 shows the major uses of some ammonium salts.

Name of compound	Use
ammonium chloride	1. Used in dry batteries 2. Used as a flux in soldering
ammonium nitrate	1. Used as a fertiliser 2. Used in making explosives
ammonium phosphate	Used as a fertiliser to provide both nitrogen and phosphorus for the soil
ammonium sulphate	Used as a fertiliser

Table 1

14.5 Making nitric acid in industry

Nitric acid (HNO_3), is another very important chemical which contains nitrogen. Millions of tons of nitric acid are made each year. Most of it is used to make ammonium nitrate, but large quantities are also used to make plastics, dyes and explosives.

In the past, nitric acid was manufactured by reacting sodium nitrate with concentrated sulphuric acid. This method is no longer used. All nitric acid is now manufactured from ammonia.

A mixture of ammonia and air is passed over a platinum catalyst at about $850\,°C$. Ammonia reacts with oxygen in the air to form nitrogen monoxide and steam:

ammonia + oxygen → nitrogen + steam
                              monoxide
$$4NH_3 + 5O_2 \rightarrow 4NO + 6H_2O$$

The mixture leaving the catalyst chamber is then cooled, and mixed with more air so that the nitrogen monoxide is changed into nitrogen dioxide:

nitrogen monoxide + oxygen → nitrogen dioxide
$$2NO + O_2 \rightarrow 2NO_2$$

The nitrogen dioxide, together with excess air is then passed into water to form nitric acid:

nitrogen + water + oxygen → nitric
dioxide   acid
$$4NO_2(g) + 2H_2O(l) + O_2(g) \rightarrow 4HNO_3(aq)$$

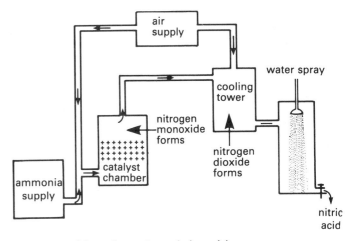

Fig 14.8 Manufacturing nitric acid

14.6 Nitric acid—the chemical

Experiment 14.2 Comparing the properties of nitric acid with hydrochloric acid

You will need: dilute nitric acid, dilute hydrochloric acid, solid sodium carbonate, magnesium ribbon, copper(II) oxide and copper turnings.

1. Place a spatula measure of sodium carbonate in a test tube. Add dilute nitric acid. Test the gas given off for carbon dioxide.

Repeat the experiment using hydrochloric acid.

2. Place a spatula measure of copper(II) oxide in a boiling tube. Add dilute nitric acid and warm the mixture. **Use safety glasses.**

Repeat the experiment using hydrochloric acid.

3. Place a 2 cm depth of dilute nitric acid in a test tube. Add magnesium ribbon. Test the gas given off for hydrogen.

Repeat the experiment using hydrochloric acid.

4. Place a 2 cm depth of nitric acid in a boiling tube. Add a few copper turnings and warm the mixture. **Use safety glasses.**

Repeat the experiment using hydrochloric acid. What are the differences between hydrochloric acid and nitric acid in these reactions?

Nitric acid behaves as an *acidic solution* and as an *oxidising agent*.

Nitric acid—the acid

Nitric acid solution shows most of the properties of acids:

It turns litmus paper red.

It reacts with carbonates forming carbon dioxide gas.

It reacts with bases forming a salt and water only.

The only way in which nitric acid behaves differently from a normal acid is: *Nitric acid does not usually produce hydrogen gas* with metals. Dilute nitric acid reacts with metals to form the nitrate salt, water and an oxide of nitrogen.

Example:

nitric + zinc → zinc + water + nitrogen
acid   nitrate   monoxide

Nitric acid—the oxidising agent

Normal acids do not react with copper, because copper is below hydrogen in the reactivity series. Nitric acid reacts with copper because it is an oxidising agent.

copper + nitric → copper(II) + nitrogen + water
acid   nitrate   dioxide

If carbon is warmed with concentrated nitric acid it gains oxygen and is oxidised to carbon dioxide.

carbon + nitric → carbon + nitrogen + water
acid   dioxide   dioxide

Concentrated nitric acid will oxidise aqueous iron(II) ions to iron(III) ions.

The nitrates

Nitric acid reacts with metals, bases and carbonates to form salts. These salts are known as nitrates. Nitrate salts have certain properties in common:

1. They all dissolve in water.
2. They are all decomposed when heated.

Although all nitrates decompose on heating, they do not decompose in the same way.

Experiment 14.3 The action of heat on nitrates

You will need sodium nitrate, zinc nitrate, lead(II) nitrate and copper(II) nitrate.

Heat a spatula measure of each nitrate in a hard-glass test tube. Note any colour changes. Test the gas given off for oxygen.

Do you notice any relationship between the action of heat of nitrates and the position of the metal in the reactivity series?

Write equations for the reactions taking place.

Most metal nitrates decompose on heating, producing brown nitrogen dioxide gas as well as oxygen.

METAL NITRATE → METAL OXIDE +
        NITROGEN DIOXIDE + OXYGEN

Example:

lead(II) → lead(II) + nitrogen + oxygen
nitrate   oxide   dioxide

$2Pb(NO_3)_2 \rightarrow 2PbO + 4NO_2 + O_2$

However the nitrates of the very reactive metals found in Group I of the Periodic Table decompose differently. They produce oxygen gas but no nitrogen dioxide on heating:

METAL → METAL + OXYGEN
NITRATE   NITRITE

Example:

sodium nitrate → sodium nitrite + oxygen

$2NaNO_3 \rightarrow 2NaNO_2 + O_2$

Questions

1. Which one of the following metals reacts with dilute nitric acid but does not react with dilute hydrochloric acid?

   A copper
   B iron
   C lithium
   D magnesium
   E zinc

2. Which one of the following set of chemicals is the most suitable for the industrial preparation of nitric acid?

   A air, ammonia and water
   B nitrogen, hydrogen, air and water
   C ammonia and hydrogen chloride
   D concentrated sulphuric acid and potassium nitrate
   E nitrogen, hydrogen and water

3. Ammonia can be obtained from ammonium sulphate by heating it:

   A alone
   B with concentrated sulphuric acid
   C with copper(II) sulphate
   D with sodium hydroxide
   E with water.

4. A gas G has the following properties:
   (i) it has a characteristic smell,
   (ii) it is very soluble in water forming an alkaline solution,
   (iii) it is colourless.
   which one of the following gases could G be?

   A ammonia
   B carbon dioxide
   C chlorine
   D hydrogen chloride
   E nitrogen

5. Which one of the following does not give off nitrogen dioxide when it is heated?

A copper(II) nitrate
B iron(III) nitrate
C lead(II) nitrate
D potassium nitrate
E zinc nitrate

6. AMMONIA, AMMONIUM CHLORIDE, AMMONIUM SULPHATE, CARBON DIOXIDE, IRON, LEAD(II) NITRATE, NITRIC ACID, PLATINUM, SODIUM NITRATE, SULPHURIC ACID.
Choose from the above:

a An acid which will not produce hydrogen gas with zinc metal.

b A solid that produces brown fumes on heating.

c A gas which will reduce hot copper(II) oxide.

d A substance which will produce oxygen on heating.

e The substance used as the catalyst in the Haber process.

f An alkaline gas.

g A substance used as a nitrogen fertiliser.

h An acid formed in the atmosphere during thunder storms.

i A substance which thermally dissociates.

7. Ammonium sulphate $(NH_4)_2SO_4$, ammonium nitrate NH_4NO_3, and urea $CO(NH_2)_2$ are three commonly used nitrogen fertilisers.

a Calculate the mass of 1 mole of each substance.

b Calculate the mass of nitrogen in 1 mole of each substance.

c Which substance contains the greatest percentage by mass of nitrogen?

d What factors (apart from percentage nitrogen) would you consider before deciding which was the best fertiliser?

8. Suppose you were given a mixture of ammonium chloride, sodium chloride and sand. How would you get a pure sample of each?

9. Nitrogen dioxide (NO_2) can be obtained by the action of heat on lead(II) nitrate. Nitrogen dioxide reacts with cold water forming nitrous acid (HNO_2) and nitric acid (HNO_3) as the only products.

a Describe what you would *observe* if lead(II) nitrate were heated until the reaction had finished and the residue left to cool.

b (i) Write the equation for the action of heat on lead(II) nitrate.
(ii) Construct the equation for the reaction between nitrogen dioxide and a cold aqueous solution of sodium hydroxide.

c Suggest why each of the following nitrates is not used for the preparation of nitrogen dioxide: (i) sodium nitrate $(NaNO_3)$, (ii) copper(II) nitrate $(Cu(NO_3)_2.3H_2O)$.

d How is nitrogen dioxide formed during the manufacture of nitric acid from ammonia?

[C]

10. Ammonium sulphate can be manufactured by absorbing ammonia gas in a suspension of calcium sulphate in water and then passing in carbon dioxide. Calcium carbonate is the other product of the reaction.

a Write an equation for the manufacture of ammonium sulphate by this method. How can ammonium sulphate be obtained from the mixture of ammonium sulphate and calcium carbonate?

b How would you prepare a dry sample of ammonium sulphate crystals from aqueous ammonia and dilute sulphuric acid?

c Give *one* commercial use of ammonium sulphate.

d When ammonium chloride is heated, thermal dissociation takes place. Explain what is meant by *thermal dissociation*. Describe clearly what you would see if ammonium chloride were heated and write an equation for the reaction.
Suggest why ammonium carbonate is less stable when heated than ammonium chloride.

[C]

11. The element phosphorus has an atomic number of 15 and a mass number of 31. Phosphorus occurs as two allotropes, a red form and a white form; only the white form is soluble in the solvent carbon disulphide. The element can combine with chlorine to form two chlorides of formulae PCl_3 and PCl_5.

a State the numbers of protons and neutrons present in the phosphorus nucleus and give a simple diagram to show the arrangement of the electrons in an atom of phosphorus.

b Explain what is meant by *allotropes* and name two other elements that exist as allotropes.

c How could a pure sample of red phosphorus be obtained from a mixture of the red and white allotropes?

d Predict the formulae of the compounds you would expect phosphorus to form with: (i) bromine, (ii) oxygen.

e The chloride PCl_3 reacts with water to form the acids H_3PO_3 and HCl and the chloride PCl_5 reacts to form the acids H_3PO_4 and HCl. Write equations for these two reactions.

[C]

12.

a Draw a labelled diagram of the apparatus you would use to prepare and collect a dry sample of ammonia gas. Write the equation for the reaction.

b Draw a diagram showing the arrangement of the electrons in a molecule of ammonia. (In your diagram, distinguish clearly between the electrons from nitrogen and those from hydrogen atoms.)

c Pure dry ammonia and oxygen were passed through the apparatus shown and the ammonia was ignited.

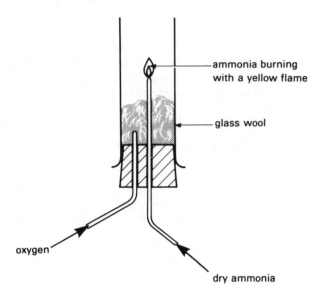

ammonia burning with a yellow flame

glass wool

oxygen

dry ammonia

(i) What is the purpose of the glass wool?

(iii) Write the equation for the combustion of ammonia in this experiment.

(iii) What would happen if the oxygen supply were replaced by an air supply?

d Under what conditions is ammonia oxidised in the manufacture of nitric acid? Give an equation for the reaction.

e Hydrazine (N_2H_4) is a more powerful reducing agent than ammonia. Suggest what products, other than nitrogen, would be formed when hydrazine is added to: (i) chlorine, (ii) a solution containing silver ions, $Ag^+(aq)$.

[C]

15 Sulphur and its compounds

15.1 Sulphur—the element

Sulphur is a yellow solid element with a melting point of about 119 °C. It can be found in nature as the element or as a wide range of sulphur containing compounds. Most supplies of sulphur are obtained from underground deposits of the element, but an increasing amount is now obtained from natural gas.

Solid sulphur is made up of molecules. There are 8 sulphur atoms in each sulphur molecule and they are arranged in a crown shaped ring.

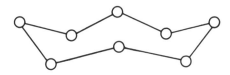

Fig 15.1 The sulphur molecule (S_8)

Solid sulphur can exist in two different crystalline forms because the crown shaped molecules can pack together in two different ways. The two crystalline forms of sulphur are *rhombic sulphur* and *monoclinic sulphur*. They are known as the *allotropes* of sulphur and can be recognised by the shape of their crystals.

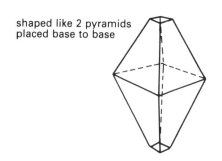

shaped like 2 pyramids placed base to base

Fig 15.2 Rhombic sulphur crystal

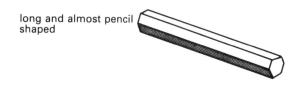

long and almost pencil shaped

Fig 15.3 Monoclinic sulphur crystal

You should remember that carbon can also exist as allotropes. Can you remember the names of the allotropes of carbon?

Making the sulphur allotropes

Rhombic sulphur: Crystals of rhombic sulphur can be made by dissolving sulphur in warm dimethyl-benzene at 40 °C. After filtering off the excess sulphur, the solution can be left to cool in an evaporating basin.

Note: Since dimethylbenzene is *both* poisonous and highly flammable, this experiment must be carried out in a fume cupboard using a water bath for heating.

Monoclinic sulphur: Crystals of monoclinic sulphur can be made by *gently* heating sulphur until it *just* melts and then pouring the liquid sulphur into a double thickness cone of filter paper. When a skin first forms on the surface of the sulphur, the filter paper can be opened out. Needle shaped crystals of monoclinic sulphur will have formed.

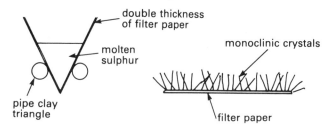

Fig 15.4 Making monoclinic sulphur

At room temperature rhombic sulphur is the more stable form. This means that any sample of monoclinic sulphur will gradually change to rhombic sulphur if left in the laboratory.

Sulphur—the chemical

Sulphur is a fairly reactive element. It will burn in air or oxygen with a blue flame to form sulphur dioxide gas:

sulphur + oxygen → sulphur dioxide
$$S + O_2 → SO_2$$

Sulphur is in the same Group of the Periodic Table as oxygen. It is therefore similar to oxygen in its chemical properties. Most metals when heated in oxygen form the metal oxide. Most metals when heated with sulphur form the metal sulphide.

Example: iron + sulphur → iron(II) sulphide
$$Fe + S → FeS$$

Uses of sulphur

Most sulphur is used to make sulphuric acid, but it does have other important uses. It can be used to harden rubber. This process is known as *vulcanising*. It is also used in matches and fireworks.

15.2 Sulphur dioxide

Sulphur dioxide is a colourless, poisonous gas with an irritating smell. It is denser than air and dissolves in water forming an acidic solution.

Making sulphur dioxide

In the laboratory sulphur dioxide is made by warming dilute sulphuric acid with sodium sulphite:

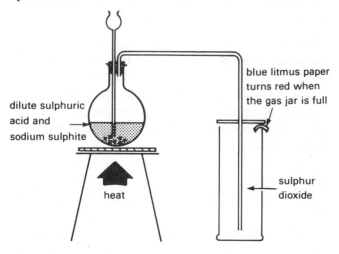

Fig 15.5 Making sulphur dioxide in the laboratory

sodium + sulphuric→ sodium + sulphur +water
sulphite   acid   sulphate   dioxide
$$Na_2SO_3 + H_2SO_4 \rightarrow Na_2SO_4 + SO_2 + H_2O$$
Because the gas is soluble in water it has to be collected by downward delivery.

Industrially, large quantities of sulphur dioxide are needed to manufacture sulphuric acid. It is made in a number of ways.

1. By burning sulphur in air. This is the major source of sulphur dioxide at present.
2. By roasting sulphide ores. Metals such as zinc and lead are found as sulphides. These sulphides have to be changed into oxides, by heating in air, before the metal can be extracted. This reaction produces useful sulphur dioxide:

zinc   + oxygen→ zinc + sulphur
sulphide   oxide   dioxide
$$2ZnS + 3O_2 \rightarrow 2ZnO + 2SO_2$$

Sulphur dioxide—the chemical

Sulphur dioxide is an acidic gas. It dissolves in water to form sulphurous acid:

sulphur + water ⇌ sulphurous
dioxide   acid
$$SO_2 + H_2O \rightleftharpoons H_2SO_3 \rightleftharpoons 2H^+ + SO_3^{2-}$$
Sulphur dioxide reacts with aqueous sodium hydroxide to give the normal salt, sodium sulphite; or the acid salt, sodium hydrogensulphite.
$$2NaOH(aq) + SO_2(g) \rightarrow Na_2SO_3(aq) + H_2O(l)$$
$$NaOH(aq) + SO_2(g) \rightarrow NaHSO_3(aq)$$

An excess of sulphur dioxide will produce the acid salt, as can be seen from the above equations.

Sulphur dioxide as a reducing agent

Sulphur dioxide, when dissolved in water, acts as a reducing agent because it is easily oxidised to sulphuric acid:

sulphurous +   oxygen   → sulphuric
  acid   (from a compound)   acid
$$H_2SO_3 + [O] \rightarrow H_2SO_4$$
Test for sulphur dioxide: it will turn potassium manganate(VII) solution colourless. It will turn potassium dichromate(VI) solution from orange to dark green.

The sulphur dioxide has reduced the manganate(VII) ion and the dichromate(VI) ion.

$$MnO_4^- \xrightarrow{\text{sulphur dioxide}} Mn^{2+}$$
(purple)   (colourless)
oxidation   oxidation
number of   number of
Mn = +7   Mn = +2

$$Cr_2O_7^{2-} \xrightarrow{\text{sulphur dioxide}} Cr^{3+}$$
(orange)   (green)
oxidation   oxidation
number of   number of
Cr = +6   Cr = +3

In both these reactions, the oxidation number of the transition element has been reduced. As the sulphur dioxide is converted to sulphuric acid, the oxidation number of sulphur has changed from +4 to +6.

It is not necessary for a compound to contain oxygen for a redox reaction to occur with sulphur dioxide. If sulphur dioxide is passed into a warm solution of iron(III) chloride, the colour of the solution changes from a yellow-brown to a pale green colour. This shows that iron(III) has been reduced to iron(II).
$$2Fe^{3+}(aq) + SO_2(g) + 2H_2O(l) \rightarrow$$
yellow-brown
$$2Fe^{2+}(aq) + 4H^+(aq) + SO_4^{2-}(aq)$$
pale green

The halogens are reduced to halide ions in the presence of sulphur dioxide:
$$SO_2 + 2H_2O + Cl_2 \rightarrow H_2SO_4 + 2HCl$$
In this reaction the oxidation number of chlorine changes from 0 (in Cl_2) to -1 (in Cl^-).

Sulphur dioxide—the polluter

Sulphur dioxide is one of the main polluting gases in industrial areas. It is formed whenever any fuel containing sulphur (coal, oil, gas) is burnt. Since it is an acidic gas it dissolves in rain water making an acidic solution. Acidic rainwater attacks limestone and many metals, causing them to corrode. Much of the recent decay of churches and cathedrals has been blamed on acidic polluting gases like sulphur dioxide in the atmosphere.

Uses of sulphur dioxide

The main use of sulphur dioxide is to manufacture sulphuric acid, but it does have several other uses:

1. It is used to bleach wood pulp during paper making.
2. It is used as a food preservative.

15.3 Sulphuric acid

You may know that sulphuric acid is used in car batteries, but did you know that sulphuric acid is one of the most important chemicals in the world? Millions and millions of tons of it are used by industry every year. It is used to make fertilisers, paints, plastics, detergents and dyes. Almost every industry has some use for it.

Making sulphuric acid

Sulphuric acid is manufactured by the *Contact process*. The raw materials are sulphur, air and water. Sulphur dioxide is produced by burning sulphur.

The sulphur dioxide is mixed with air and passed over a catalyst of vanadium(V) oxide at 450 °C. The sulphur dioxide reacts with oxygen in the air to form sulphur trioxide:

sulphur + oxygen $\xrightleftharpoons[\text{vanadium(V) oxide}]{\text{450 °C}}$ sulphur
dioxide   trioxide
$$2SO_2 + O_2 \rightleftharpoons 2SO_3$$

Sulphur trioxide reacts with water to form sulphuric acid:

sulphur trioxide + water → sulphuric acid
$$SO_3 + H_2O \rightarrow H_2SO_4$$

Unfortunately, if sulphur trioxide is passed into water the sulphuric acid is formed as a fine mist. This mist is difficult to condense. Instead, sulphur trioxide is dissolved in 98% sulphuric acid. A controlled amount of water is added at the same time so that the acid is kept at 98% sulphuric acid.

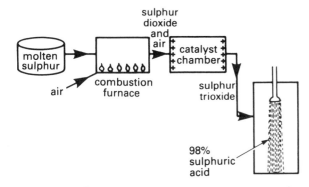

Fig 15.6 The Contact process

The burning of sulphur and the changing of sulphur dioxide into sulphur trioxide are exothermic reactions. They produce a great deal of heat.

Sulphuric acid—the chemical

Dilute sulphuric acid behaves like a typical acid. It turns litmus paper red. It reacts with most metals producing hydrogen gas.

Example:

magnesium + sulphuric → magnesium + hydrogen
acid   sulphate
$$Mg(s) + H_2SO_4(aq) \rightarrow MgSO_4(aq) + H_2(g)$$

It also reacts with metal carbonates producing carbon dioxide gas.

Example:

copper(II) + sulphuric → copper(II) + carbon + water
carbonate   acid   sulphate   dioxide
$$CuCO_3(s) + H_2SO_4(aq) \rightarrow CuSO_4(aq) + CO_2(g) + H_2O(l)$$

It also reacts with bases to form a salt and water only.

Example:

magnesium + sulphuric → magnesium + water
oxide   acid   sulphate
$$MgO(s) + H_2SO_4(aq) \rightarrow MgSO_4(aq) + H_2O(l)$$

While dilute sulphuric acid behaves as a typical acid, concentrated sulphuric acid has some different properties.

1. *As a drying agent*

Concentrated sulphuric acid has a very strong attraction for water. It is therefore used to dry many gases. The gases are dried by bubbling them through concentrated sulphuric acid. Ammonia and ethene are the only two common gases that cannot be dried using concentrated sulphuric acid. They react with it.

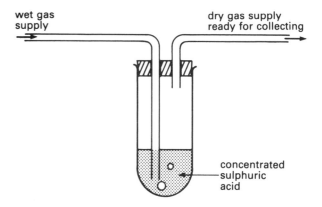

Fig 15.7 Using concentrated sulphuric acid to dry gases

2. *As a dehydrating agent*

So strong is the attraction of concentrated sulphuric acid for water that it will even remove water from some compounds.

If blue copper(II) sulphate crystals are dropped into concentrated sulphuric acid they turn white as they lose their water of crystallisation:

blue copper(II) → white anhydrous + water
sulphate crystals   copper(II) sulphate
$$CuSO_4.5H_2O \rightarrow CuSO_4 + 5H_2O$$

With some compounds such as sugar it will remove hydrogen and oxygen from the compound as water:

sucrose → carbon + water
$$C_{12}H_{22}O_{11} \rightarrow 12C + 11H_2O$$

When concentrated sulphuric acid breaks down a compound, forming water, it is behaving as a *dehydrating* agent. The process is known as *dehydration*. Concentrated sulphuric acid also dehydrates skin and flesh. It can cause very serious and painful wounds.

3. Displacement of volatile acids

As concentrated sulphuric acid has a high boiling point (338 °C), it can be used to produce *volatile* acids (acids with low boiling points) from their salts. Two common volatile acids are: hydrogen chloride, boiling point −85 °C, and nitric acid, boiling point 86 °C.

Example:

$$H_2SO_4(l) + NaCl(s) \longrightarrow NaHSO_4(s) + HCl(g)$$

$$H_2SO_4(l) + NaNO_3(s) \xrightarrow{\text{heat}} NaHSO_4(s) + HNO_3(g)$$

Questions

1. Which one of the following sets of conditions, together with a catalyst, is used in the Contact process for the manufacture of sulphuric acid?

A a pressure of 200 atmospheres and a temperature of 200 °C

B a pressure of 200 atmospheres and a temperature of 400 °C

C atmospheric pressure and room temperature

D atmospheric pressure and a temperature of 500 °C

E a pressure of 500 atmospheres and a temperature of 2000 °C

2. When concentrated sulphuric acid reacts with sodium chloride solid at room temperature and pressure the only products are:

A an acid salt and a normal salt

B a salt, a base and water

C an acid salt and an acidic gas

D a normal salt and water

E a base and an acidic gas.

3. Selenium, Se, is in the same group of the Periodic Table as sulphur. What is the formula of magnesium selenide?

A Mg_2Se

B $MgSe$

C $MgSe_2$

D $MgSeO_3$

E $MgSeO_4$

4. Sulphur dioxide gas is produced by the reaction between which pair of the following ions?

A $SO_3^{2-}(aq)$ and $H^+(aq)$

B $SO_4^{2-}(aq)$ and $H^+(aq)$

C $SO_3^{2-}(aq)$ and $OH^-(aq)$

D $Cu^{2+}(aq)$ and $SO_4^{2-}(aq)$

E $SO_3^{2-}(aq)$ and $Cl^-(aq)$

5. Which one of the following gases cannot be dried using concentrated sulphuric acid?

A ammonia

B chlorine

C hydrogen

D hydrogen chloride

E sulphur dioxide

6. Monoclinic sulphur and rhombic sulphur are the allotropes of sulphur.

a What are allotropes?

b Name another element that has allotropes.

c If you were shown crystals of rhombic sulphur and monoclinic sulphur how would you know which was which?

7.

a How can sulphur be changed into sulphur dioxide? Write an equation for the reaction.

b Name two compounds which when mixed together at room temperature react to form sulphur dioxide. Write an equation for the reaction.

c How would you test a gas to show that it was sulphur dioxide?

d Sulphur dioxide is one of the main gases that pollute the atmosphere. It is responsible for much of the decay of limestone buildings. What property of sulphur dioxide makes it attack these buildings?

8. Write word equations and symbol equations for the reactions of dilute sulphuric acid with the following:

a copper(II) carbonate   c magnesium

b sodium hydroxide   d zinc oxide

9. If burning magnesium is lowered into a gas jar of sulphur dioxide a white powder P and a pale yellow powder Q are formed. P reacts with dilute sulphuric acid to form R and water. Q does not react with dilute sulphuric acid. Q burns in air to reform sulphur dioxide.

a Identify P, Q and R.

b Write equations for the following reactions:
   (i) the reaction between sulphur dioxide and magnesium
   (ii) the reaction between P and dilute sulphuric acid
   (iii) the burning of Q.

10.

a Concentrated sulphuric acid is used to dry gases. Name one gas which cannot be dried using sulphuric acid.

b Give an example of a reaction in which concentrated sulphuric acid behaves as a dehydrating agent.

c What is the difference between drying and dehydrating?

11.

a Describe the observations, if any, you would make when dilute sulphuric acid reacts with:
   (i) zinc metal
   (ii) sodium hydroxide solution
   (iii) copper(II) carbonate
   (iv) blue litmus paper
   Equations for the reactions are not required.

b Giving full practical details, state clearly how

you would obtain pure, dry crystals of copper-(II) sulphate ($CuSO_4.5H_2O$) from a sample of the crystals contaminated with copper(II) carbonate.

c Oxides of elements may be classified as acidic, basic, neutral or amphoteric.
Give the *name* of one example of each of these types of oxide.

[C]

12. Sulphuric acid can be manufactured by the Contact process.

a (i) Give the names of the two gases which react together in the process.
(ii) Give the names of the raw materials which are used as sources of these gases.
(iii) Give the name of the product of the reaction between the two gases in (i).
(iv) Write an equation for the reaction.
(v) How is the product of this reaction converted into sulphuric acid?

b In the Contact process a catalyst is used.
(i) Why is a catalyst used?
(ii) Give the name of a suitable catalyst.

c Sulphuric acid can be converted into a fertilizer. Give the chemical name of *one* such fertilizer.

[L]

13. The element selenium (symbol Se) is placed in Group VI of the Periodic Table, immediately below sulphur.

a Which would you expect selenium to be, a metal or a non-metal?

b When heated in oxygen, selenium forms a solid oxide which dissolves in water.
(i) What would you expect the formula of this oxide to be?
(ii) Write an equation for its reaction with water.
(iii) How would you expect the solution to react with Universal indicator paper?

c Selenium forms a gaseous compound with hydrogen called hydrogen selenide, containing 97·53% by mass of selenium.
(i) How many moles of selenium atoms are present in 100 g of this gas?
(ii) How many moles of hydrogen atoms are present in 100 g of this gas?
(iii) How many moles of hydrogen atoms combine with 1 mole of selenium atoms?
(iv) On this evidence, what is the formula of hydrogen selenide?

d Hydrogen selenide decomposes into its elements on heating. What volume of hydrogen, measured at the same temperature and pressure, would you expect to obtain if 25 cm^3 of hydrogen selenide were to be completely decomposed?

[C]

14.
a Describe briefly, with equations, the manufacture of sulphuric acid from sulphur by the Contact process. State one source of the sulphur which is used.

b Using a different substance in each case, choose one of the following substances (copper, magnesium oxide, sugar, zinc) to illustrate the reaction of sulphuric acid as:
(i) an acid, (ii) a dehydrating agent, (iii) an oxidising agent.

Your answer should include (a) the reaction conditions, (b) a statement of what would be observed and (c) either an equation for, *or* the names of the products formed in, each reaction.

[JMB]

16 Chlorine and the other halogens

The element chlorine is a greenish–yellow gas. It is very reactive and even reacts with gold. Because it is so reactive, there is no free chlorine in the atmosphere. Instead, chlorine is found in compounds. The most plentiful compound of chlorine is sodium chloride. It is found in large underground deposits as rock salt and also dissolved in sea water. No wonder it is known as common salt. Sodium chloride is used to make a number of useful chemicals that are not found naturally.

16.1 Hydrogen chloride

When concentrated sulphuric acid is added to sodium chloride a reaction takes place and a colourless gas is formed:

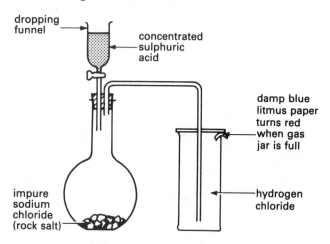

Fig 16.1 Making hydrogen chloride in the laboratory

sodium + concentrated → sodium + hydrogen
chloride   sulphuric   hydrogen   chloride
           acid   sulphate

$$NaCl(s) + H_2SO_4(l) \rightarrow NaHSO_4(s) + HCl(g)$$

Hydrogen chloride can be made in the laboratory using the apparatus shown in Fig 16.1. The gas has to be collected by downward delivery as it is more dense than air and very soluble in water.

Hydrogen chloride—the chemical

Hydrogen chloride is a colourless gas with a sharp irritating smell. It does not burn and it does not allow other substances to burn in it. Hydrogen chloride reacts with ammonia gas to form a thick white smoke of ammonium chloride:

ammonia + hydrogen → ammonium
          chloride   chloride

$$NH_3(g) + HCl(g) \rightarrow NH_4Cl(s)$$

This reaction is often used as a test for hydrogen chloride. Concentrated ammonia solution can be used as a source of ammonia.

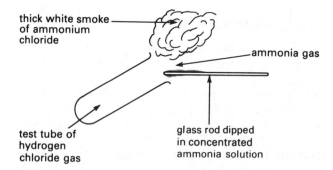

Fig 16.2 Testing for hydrogen chloride

When hydrogen chloride comes into contact with moist air it forms a white mist. We say that hydrogen chloride fumes in moist air. Hydrogen chloride is in fact very soluble in water. Special precautions have to be taken when making a solution of hydrogen chloride in water.

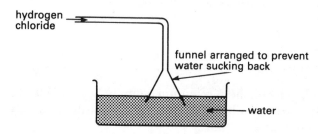

Fig 16.3 Making a solution of hydrogen chloride in water

Hydrogen chloride is a covalent substance, but when it dissolves in water it reacts with the water forming ions. The solution formed is known as *hydrochloric acid.*

hydrogen + water → hydrochloric
chloride   acid

$$HCl + H_2O \rightarrow H_3O^+ + Cl^-$$

Dry hydrogen chloride does not have the properties of an acid. It does not affect *dry* litmus paper and does not produce carbon dioxide from carbonates. However, hydrogen chloride will react with some metals producing hydrogen gas. This can be demonstrated using the apparatus shown in Fig 16.4.

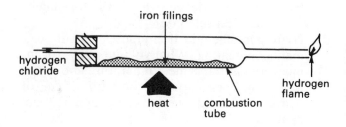

Fig 16.4 Reacting iron with hydrogen chloride

iron + hydrogen → iron(II) + hydrogen
     chloride   chloride
$$Fe(s) + 2HCl(g) → FeCl_2(s) + H_2(g)$$
The hydrogen formed can be burnt at the end of the combustion tube.

16.2 Hydrochloric acid

We have seen that hydrochloric acid is a solution of hydrogen chloride in water. If the solution is saturated with hydrogen chloride it is known as concentrated hydrochloric acid. Concentrated hydrochloric acid contains about 36% hydrogen chloride by mass.

Hydrochloric acid—the chemical

Hydrochloric acid behaves as a typical acid when concentrated and when dilute. It turns blue litmus paper red. It reacts with most metals producing hydrogen gas:

Example:

magnesium + hydrochloric → magnesium + hydrogen
          acid   chloride
$$Mg(s) + 2HCl(aq) → MgCl_2(aq) + H_2(g)$$
It reacts with carbonates producing carbon dioxide gas.

Example:

calcium + hydrochloric → calcium + carbon + water
carbonate   acid   chloride dioxide
$$CaCO_3(s) + 2HCl(aq) → CaCl_2(aq) + CO_2(g) + H_2O(l)$$
It also reacts with bases to form salts and water only.

Example:

copper(II) + hydrochloric → copper(II) + water
  oxide   acid   chloride
$$CuO(s) + 2HCl(aq) → CuCl_2(aq) + H_2O(l)$$

Apart from its acidic properties, hydrochloric acid can be oxidised to chlorine using manganese(IV) oxide.

Example:

concentrated + manganese → manganese
hydrochloric (IV) oxide   (II)chloride
  acid

                   + water + chlorine
$$4HCl(aq) + MnO_2(s) → MnCl_2(aq)$$
$$+ 2H_2O(l) + Cl_2(g)$$

Note that manganese(IV) oxide is acting as an oxidising agent (Fig. 16.5).

16.3 Chlorine

Chlorine is made in the laboratory by oxidising hydrochloric acid using either potassium manganate(VII) or manganese(IV) oxide. If potassium manganate(VII) is used, no heat is required.

Chlorine is collected by downward delivery because it is soluble in water and denser than air.

Chlorine is also formed when metallic chlorides are electrolysed (Section 10.3).

Chlorine is manufactured industrially by electrolysis of molten sodium chloride (Section 11.2) and of aqueous sodium chloride (Section 10.3).

Chlorine—the chemical

The element chlorine exists as Cl_2 molecules. Each chlorine atom has seven electrons in its outermost shell. By two atoms sharing electrons the chlorine molecule is formed.

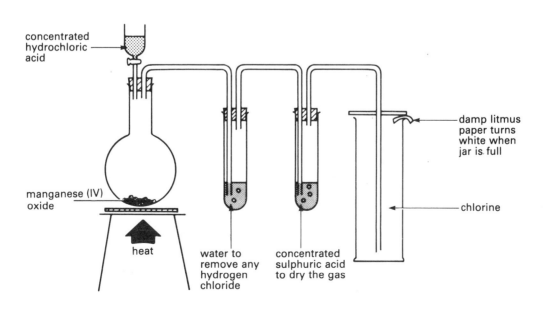

Fig 16.5 Preparing chlorine in the laboratory

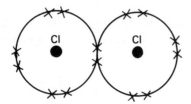

Fig 16.6 The chlorine molecule

Chlorine has a choking, irritating smell and is extremely poisonous. Its poisonous nature and the fact that it is more than twice as dense as air help to explain why it was used as a weapon during World War 1.

Chlorine does not burn but it is a very reactive element. It reacts with most elements, except carbon, to form the chloride of the element.

With metallic elements, it reacts to form ionic salts, for example, sodium chloride. With non-metallic elements such as phosphorus, it reacts to form covalent chlorides, such as phosphorus trichloride.

Chlorine is moderately soluble in water forming a solution known as *chlorine water*. Dissolved chlorine reacts with water to form a mixture of hydrochloric and chloric(I) acids.

$$\text{chlorine} + \text{water} \rightleftharpoons \text{hydrochloric} + \text{chloric(I)}$$
$$\text{acid} \qquad \text{acid}$$
$$Cl_2 + H_2O \rightleftharpoons HCl + HClO$$

Chlorine reacts with dilute sodium hydroxide to form a mixture of sodium chlorate(I) (NaClO) and sodium chloride.

By looking at the reaction of chlorine with water, work out the equation for the reaction of sodium hydroxide with chlorine.

Chlorine gas is an oxidising agent because it is easily reduced to chloride ions.

It oxidises sulphites to sulphates, iron(II) compounds to iron(III) compounds, bromide ions to bromine and iodide ions to iodine.

$$Cl_2 + SO_3^{2-} + H_2O \rightarrow SO_4^{2-} + 2Cl^- + 2H^+$$
$$2Fe^{2+} + Cl_2 \rightarrow 2Fe^{3+} + 2Cl^-$$

It produces iron(III) chloride when reacted with iron whereas hydrogen chloride produces iron(II) chloride when reacted with iron.

$$2Fe + 3Cl_2 \rightarrow 2FeCl_3$$
$$Fe + 2HCl \rightarrow FeCl_2 + H_2$$

Testing for chlorine

Chlorine gas will turn damp blue litmus paper red and then bleach it. This is the usual test for chlorine.

Uses of chlorine

1. Chlorine is used to sterilise domestic water supplies. After water from reservoirs has been filtered to remove any solids, it may still contain harmful germs. These are killed by passing chlorine into our water supply before it reaches our homes. The amount of chlorine used must be very carefully controlled. It must be enough to kill all the germs and yet not enough to give the water an unpleasant taste. Certainly there must not be enough chlorine in the water to harm the people using it.
2. Chlorine is used in swimming pools to kill harmful germs. It is used in much higher concentrations than in domestic water supplies. The chlorine taste makes the water most unpleasant if you swallow some while swimming.
3. Chlorine is used in the manufacture of plastics. The main plastic which contains chlorine is polyvinylchloride (PVC).
4. Chlorine is used to make a number of industrial solvents such as trichloroethane, which is widely used for degreasing metals.
5. Chlorine is used to make domestic bleaches. Most domestic bleaches are made by dissolving chlorine gas in sodium hydroxide solution:

sodium + chlorine→ sodium + sodium + water
hydroxide   chloride   chlorate(I)
$$2NaOH + Cl_2 \rightarrow NaCl + NaOCl + H_2O$$

16.4 The halogens

The halogens are the elements in Group VII of the Periodic Table. They include fluorine, chlorine, bromine and iodine.

Name of halogen	Formula	Appearance at room temperature
fluorine	F_2	pale yellow gas
chlorine	Cl_2	greenish–yellow gas
bromine	Br_2	brown liquid (easily vaporised)
iodine	I_2	shiny grey solid (easily vaporised)

Table 1 The halogens

Experiment 16.1 Reactions of the halogens

You will need: aqueous potassium chloride, aqueous potassium bromide, aqueous potassium iodide, chlorine water, bromine water and a solution of iodine in potassium iodide and 1,1,1-trichloroethane.

1. Add chlorine water to aqueous potassium bromide followed by a few drops of 1,1,1-trichloroethane. Shake. Note any colour changes.

Repeat the experiment by adding bromine water and chlorine water to potassium iodide, and bromine water to potassium iodide.

Do bromine water and iodine solution react with potassium chloride solution?

2. Place about 1 cm depth of aqueous potassium chloride in a test tube and add a few drops of aqueous silver nitrate. Note the colour of any precipitate formed.

Add excess aqueous ammonia. Is the precipitate soluble in excess ammonia?

Repeat test 2 with potassium bromide and potassium iodide.

How could you distinguish between chloride, bromide and iodide ions?

3. Test chlorine water, bromine water and the iodine solution with acidified iron(II) sulphate to see if they are oxidising agents.

All the halogens exist as *diatomic molecules*. This means that there are two atoms in each molecule of the element. They have similar chemical properties even though they differ greatly in appearance. Fluorine is the most reactive and they become less reactive going down the group:

Fluorine	Chlorine	Bromine	Iodine
most			**least**
reactive	decreasing reactivity		**reactive**

You may remember that a more reactive metal will displace a less reactive metal from a solution of one of its salts. In the same way, a more reactive halogen will displace a less reactive halogen from a solution of one of its salts.

If chlorine is bubbled into a solution of potassium bromide, bromine is formed:

Example:

chlorine + potassium → bromine + potassium
                  bromide   chloride

$$Cl_2 + 2KBr \rightarrow Br_2 + 2KCl$$
$$Cl_2(g) + 2Br^-(aq) \rightarrow Br_2(l) + 2Cl^-(aq)$$

Uses of the halogens and their compounds

Fluorine is used in the manufacture of inert plastics such as Teflon which is used for non-stick saucepans.

Bromine is used to manufacture silver bromide which is light sensitive and used in photographic films.

Iodine is used as a mild antiseptic when dissolved in ethanol. Silver iodide is used in high speed films as it is more light sensitive than silver bromide.

Questions

1. To obtain pure chlorine from the reaction between hot concentrated hydrochloric acid with manganese(IV) oxide the impure chlorine is passed through two liquids, as shown in the diagram:

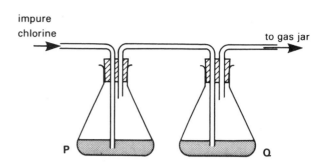

What are the most suitable liquids to put into P and Q?

	P	Q
A	lime water	water
B	water	concentrated sulphuric acid
C	sodium hydroxide solution	water
D	sodium hydroxide solution	concentrated sulphuric acid
E	hydrochloric acid	lime water

2. Which one of the following reactions produces hydrogen chloride gas?

A pass chlorine through water
B burn hydrogen in chlorine
C electrolyse aqueous sodium chloride
D react sodium chloride with dilute sulphuric acid
E add aqueous silver nitrate to aqueous sodium chloride

3. Which one of the following elements does not react with chlorine?

A carbon
B copper
C hydrogen
D iron
E sodium

4. Which one of the following forms a colourless solution which reacts with aqueous silver nitrate in the presence of dilute nitric acid, to give a yellow precipitate insoluble in excess aqueous ammonia?

A sodium iodide
B copper(II) bromide
C iron(III) nitrate
D lead(II) chloride
E magnesium carbonate

5. When chlorine is passed into a solution of iron(II) chloride the solution changes from a green solution to a brown solution. This reaction is an example of:

A decomposition
B displacement
C hydrolysis
D neutralisation
E redox.

6. BROMINE, CHLORINE, HYDROGEN CHLORIDE, IODINE, SILVER BROMIDE, SODIUM CHLORIDE.
Choose from the above list:

a A halogen which is a liquid at room temperature.

b A green gas.

c A chemical found in sea water but not in fresh water.

d A substance decomposed by sunlight.

e A colourless gas that fumes in damp air.

f A substance used to preserve food.

g A gas that dissolves in water to form two different acids.

h A solid element in Group VII of the Periodic Table.

i A substance that bleaches damp litmus paper.

7. Fluorine, atomic number 9, exists as F_2 molecules. Draw a diagram to show the arrangement of electrons in a fluorine molecule.

8. Write word equations and symbol equations for the reactions of dilute hydrochloric acid with the following:

a calcium carbonate

b magnesium

c sodium hydroxide

d copper(II) oxide.

9. A white crystalline solid K reacts with concentrated sulphuric acid to form a colourless gas L. L is an acidic gas and forms white fumes in moist air. A concentrated solution of L in water reacts with manganese(IV) oxide to form a green gas M.

When a piece of hot sodium metal is placed in gas M the white solid K is reformed.

Identify K, L and M giving your reasons.

10. Hydrogen burns in chlorine to form hydrogen chloride.

a Write an equation for the reaction.

b Hydrogen chloride is extremely soluble in water. Explain how you would safely make a solution of hydrogen chloride in water in the laboratory. Include a diagram of the apparatus you would use.

c How would you show that a solution of hydrogen chloride in water:
  (i) was acidic
  (ii) contained ions?

d What is the usual name for a solution of hydrogen chloride in water?

11. Phosphorus is in the same Group of the Periodic Table as nitrogen. Phosphorus trichloride, PCl_3, is a liquid having a boiling point of 76°C. It can be prepared by passing pure dry chlorine over heated phosphorus. Phosphorus trichloride reacts vigorously with water according to the equation

$$PCl_3(l) + 3H_2O(l) \rightarrow H_3PO_3(aq) + 3HCl(g)$$

a Draw a labelled diagram of the apparatus you would use to *prepare* and *collect* a sample of phosphorus trichloride, given phosphorus and a supply of chlorine.

b Write the equation for the reaction between phosphorus and chlorine. Why is it necessary for *dry* chlorine to be used in this reaction?

c Suggest why the experiment should be carried out in a fume cupboard.

d What mass of phosphorus trichloride would be obtained from 0·1 mole of phosphorus atoms?

e The arrangement of the electrons in a molecule of water can be represented as

$$H : \overset{\times\times}{\underset{\times\times}{O}} : H$$

In a similar way show the arrangement of outer shell electrons in a molecule of phosphorus trichloride.

f What would you observe if phosphorus trichloride were added to:
(i) litmus solution, (ii) aqueous silver nitrate?
[C]

12. Anhydrous iron(III) chloride can be prepared by the reaction between hot iron and dry chlorine, freed from hydrogen chloride. The iron(III) chloride sublimes from the reaction tube and can be collected as black crystals in a cooled receiver.

The chlorine can be prepared by the reaction between hot concentrated hydrochloric acid and manganese(IV) oxide; the other products of this reaction are manganese(II) chloride and water.

a Give diagrams of the apparatus required: (i) to prepare pure, dry chlorine, (ii) to prepare from pure dry chlorine a sample of anhydrous iron(III) chloride.
Construct the equations for:
 (i) the preparation of the chlorine,
 (ii) the reaction between the chlorine and the iron.

b Why is it important that the chlorine is dry?

c The iron(III) chloride sublimes from the reaction tube. What does this fact suggest concerning the bonding present in this compound? Give brief reasons for your answer.

d In this reaction the chlorine oxidizes the iron. Justify the use of the word "oxidizes" in this case.

e Name a reagent which will reduce aqueous iron(III) chloride to aqueous iron(II) chloride and write an equation for the reduction.
[C]

13. Bromine is a very corrosive liquid boiling at 59°C. It is a member of the halogen group of elements and closely resembles chlorine in its chemical properties. 0·8 g of bromine vapour occupies a volume of $112\,cm^3$, the volume being corrected to s.t.p.

a Name *two* members of the halogen group other than chlorine and bromine.

b How many valency electrons (outer shell electrons) are present in the atom of a halogen element?

c Calculate from the information above the relative molecular mass and atomicity of bromine vapour.

d Write the formulae for hydrogen bromide and the compound you would expect to be formed on mixing hydrogen bromide with ammonia.

e Bromine can be prepared by heating a mixture of sodium bromide with manganese(IV) oxide (MnO_2) and concentrated sulphuric acid. Draw a diagram of the apparatus you would use to carry out this preparation and to collect a sample of liquid bromine.

Why is manganese(IV) oxide used in this preparation?

f What would you expect to see if you passed chlorine into a solution of sodium bromide and then heated the liquid?

Write an *ionic* equation for this reaction.

[C]

17 Carbon and its compounds

Suppose you saw the following advert in a local paper:

FOR SALE 100 g of pure carbon. $50 or nearest offer.

Would you think it was very cheap or very expensive? You could be right whichever you think, as there are two forms of pure carbon—*diamond* and *graphite*. 100 g of diamond for only $50 would be an amazing bargain. $50 for 100 g of graphite is far too expensive.

17.1 Allotropy of carbon

It seems very surprising that two substances as different as diamond and graphite can be different forms of the *same* element. Table 1 shows how different their properties are.

Properties of diamond	Properties of graphite
1. Diamond is a colourless, transparent, glass-like solid.	**1.** Graphite is a dark grey, slightly shiny solid.
2. Diamond is a very hard substance that is used to cut glass.	**2.** Graphite is a soft substance with a slightly soapy feel.
3. Diamond does not conduct electricity.	**3.** Graphite conducts electricity.

Table 1

Even though diamond and graphite are so different, they can both be shown to be forms of pure carbon by burning.

Equal masses of diamond and graphite will burn in oxygen to form equal masses of carbon dioxide as the only product.

carbon + oxygen → carbon dioxide

$$C(s) + O_2(g) \rightarrow CO_2(g)$$

Chemists say that diamond and graphite are *allotropes* of carbon.

Diamond and graphite are so different because the carbon atoms are arranged differently in each form.

In diamond each carbon atom is joined to four other carbon atoms by *strong covalent bonds*. In this way a rigid, three dimensional structure is built up. This explains why diamond is such a hard substance. In the diamond structure all the outer shell electrons are used for bonding.

In graphite, each carbon atom is joined to three other carbon atoms by *strong covalent bonds*. In this way, strong, rigid, flat layers are formed. Each carbon atom has one outer shell electron which is not used to form covalent bonds. These 'spare' electrons can move along the layers, thus allowing the conduction of electricity.

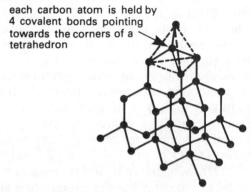

each carbon atom is held by 4 covalent bonds pointing towards the corners of a tetrahedron

Fig 17.1 The arrangement of carbon atoms in diamond

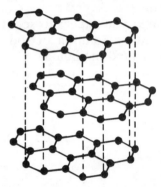

Fig 17.2 The arrangement of carbon atoms in graphite

In the graphite structure the carbon atoms are firmly held in each layer but there is little attraction between layers. The layers can easily slide over each other. When you write with a pencil, the pressure of your hand is enough to separate the layers in the graphite pencil lead. A trail of graphite layers is left on the paper recording what you write.

17.2 Making carbon for industry

Industry uses an enormous amount of carbon. It uses diamond for cutting and polishing. It uses graphite as a lubricant and an electrical conductor.

However, most of the carbon needed for industry is used in chemical reactions: reactions such as getting iron from iron ore in the blast furnace. The carbon needed for large scale chemical reactions does not need to be completely pure, but it does need to be cheap and readily available. *Coke* is the form of impure carbon most widely used in industry and is made by heating coal in the absence of air.

Coke is used industrially as a cheap reducing agent. It is used to produce metals from their oxide ores. Iron and zinc are produced in this way.

17.3 Carbon dioxide

One of the most important compounds of carbon is carbon dioxide. Normal air contains only 0·03% carbon dioxide but the air we breathe out contains far more; up to 4%. It is perhaps surprising that with so many animals breathing carbon dioxide into the atmosphere, its concentration does not build up. The main reason for this is *photosynthesis* (see Section 13.4)

In this process plants remove carbon dioxide from the atmosphere by converting it into carbohydrates.

These plants either:

1. decay and eventually become fuels such as coal (Section 7.1) which are then burned releasing carbon dioxide back into the air, or

2. they are eaten by animals or man as food.

In the process of respiration in animals and man, the carbohydrates are oxidised to carbon dioxide and water. This releases carbon dioxide back into the atmosphere. This cycle of processes is called the *carbon cycle.*

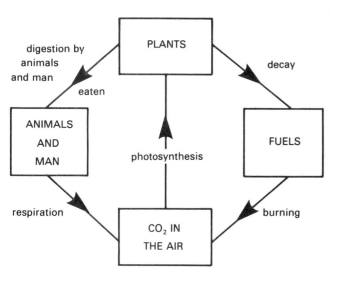

Fig 17.3 The carbon cycle

Making carbon dioxide

Carbon dioxide is usually made in the laboratory by reacting calcium carbonate with dilute hydrochloric acid. Any carbonate could be used, but since calcium carbonate is so plentiful it is the obvious choice. Calcium carbonate is found naturally as limestone, chalk and marble.

calcium carbonate	+	hydrochloric acid	→	calcium chloride	+	carbon dioxide	+	water
$CaCO_3(s)$	+	$2HCl(aq)$	→	$CaCl_2(aq)$	+	$CO_2(g)$	+	$H_2O(l)$

Because carbon dioxide is only slightly soluble in water it can be collected over water. It is also far denser than air and so can also be collected by downward delivery as shown in Fig. 17.5.

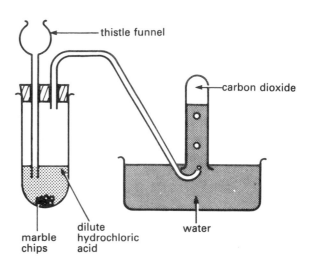

Fig 17.4 Preparation of carbon dioxide

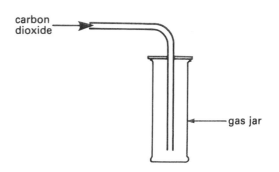

Fig. 17.5 Collecting carbon dioxide by downward delivery

Chemical properties of carbon dioxide

1. *Carbon dioxide; the acid*

Carbon dioxide is not very soluble in water. 1000 cm³ of water will only dissolve about 90 cm³ of carbon dioxide at room temperature and atmospheric pressure. When carbon dioxide dissolves in water an acidic solution is formed. This solution is known as carbonic acid solution:

water + carbon ⇌ carbonic
          dioxide   acid
$$H_2O + CO_2 \rightleftharpoons H_2CO_3 \rightleftharpoons 2H^+ + CO_3^{2-}$$

Carbonic acid is a very weak acid, but it will turn litmus paper red and Universal indicator orange or yellow. Very often, distilled water kept in laboratories becomes acidic, because it dissolves carbon dioxide from the atmosphere. When this happens, the water has to be boiled to drive out all the carbon dioxide making neutral water once again.

Because it is an acidic gas, carbon dioxide reacts with alkalis to form a salt and water only.

Example:

sodium + carbon → sodium + water
hydroxide   dioxide   carbonate
$$2NaOH + CO_2 \rightarrow Na_2CO_3 + H_2O$$

Excess of carbon dioxide produces the acid salt, sodium hydrogencarbonate ($NaHCO_3$).

The similar reaction between carbon dioxide and lime water (calcium hydroxide) is used as a test for carbon dioxide. *Carbon dioxide will turn lime water milky.*

carbon + lime → calcium + water
dioxide water carbonate
$$CO_2(g) + Ca(OH)_2(aq) \rightarrow CaCO_3(s) + H_2O(l)$$

If excess carbon dioxide is passed through the solution, the milkiness disappears because the soluble acid salt, calcium hydrogencarbonate, is formed.

calcium + water + carbon → calcium hydrogen
carbonate dioxide carbonate
$$CaCO_3(s) + H_2O(l) + CO_2(g) \rightarrow Ca(HCO_3)_2(aq)$$

If the solution of calcium hydrogencarbonate is boiled, it decomposes to give a precipitate of calcium carbonate, carbon dioxide and water.

Uses of carbon dioxide

Carbon dioxide is a colourless gas with no smell. It is used in making 'fizzy' drinks. As most substances will not burn in carbon dioxide, it is widely used in fire extinguishers. If carbon dioxide gas is cooled down at atmospheric pressure it changes directly from a gas into a solid. This solid is known as *dry ice.*

17.4 Carbonates and hydrogencarbonates

All carbonates are solids. You have seen in Table 9 (Chapter 9) that Group I carbonates are soluble in water; all the other carbonates are insoluble.

Only the hydrogencarbonates of Group I metals can be obtained as solids. The hydrogencarbonates of Group II metals are found only in solution.

Experiment 17.1 The reactions of carbonates

You will need: sodium carbonate, calcium carbonate, copper(II) carbonate and zinc carbonate.

1. Heat a spatula measure of each carbonate strongly in a hard-glass test-tube. Note any colour changes. Test the gas given off for carbon dioxide.

2. Add dilute nitric acid to a spatula measure of each carbonate in a test-tube. Test the gas given off for carbon dioxide.

Do you see any relationship between the action of heat on a metal carbonate and the position of the metal in the reactivity series?

Write equations for all the reactions taking place.

Experiment 17.2 The reactions of hydrogen-carbonates

You will need: sodium hydrogencarbonate solid and a solution of calcium hydrogencarbonate in water.

Repeat both the tests in Experiment 17.1. Heat the calcium hydrogencarbonate solution to boiling point. Use a 3 cm depth of the solution in a boiling tube. **Wear safety glasses.**

How could you distinguish between solid sodium hydrogencarbonate and sodium carbonate?

Carbonates of metals high in the reactivity series do not decompose. All other carbonates form the oxides and carbon dioxide.

Example:

lead(II) → lead(II) + carbon
carbonate oxide dioxide
$$PbCO_3 \rightarrow PbO + CO_2$$

Hydrogencarbonates decompose very easily when heated.

Example:

potassium → potassium + water + carbon
hydrogencarbonate carbonate dioxide
$$2KHCO_3 \rightarrow K_2CO_3 + H_2O + CO_2$$

Carbonates and hydrogencarbonates react with acids to form carbon dioxide, water and a salt.

magnesium + hydrochloric → carbon + water + magnesium
carbonate acid dioxide chloride
$$MgCO_3 + 2HCl \rightarrow CO_2 + H_2O + MgCl_2$$

Calcium oxide (lime)

Industry produces a large amount of acidic waste material. This waste must not be thrown into rivers, lakes and seas because it would destroy fish and plant life. The waste is neutralised by reacting it with lime.

Lime can be treated with water and added to soil that has a pH value that is too low to grow the crop needed.

Lime is manufactured by heating limestone at about 1000 °C for several hours.

calcium carbonate → calcium oxide + carbon
(limestone) (lime) dioxide
$$CaCO_3 \rightarrow CaO + CO_2$$

The carbon dioxide is allowed to escape into the atmosphere.

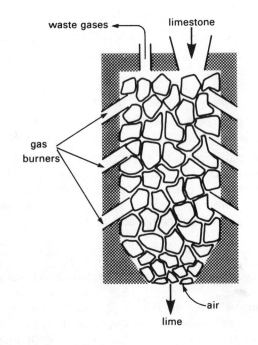

Fig 17.6 A lime kiln

114

17.5 Hard and soft water

Hard water is water that forms a scum with soap (see Section 18.7); soft water does not. The so-called hardness is due to the presence of dissolved calcium or magnesium compounds.

Rain water is acidic because it dissolves carbon dioxide from the atmosphere. This acidic rain water forms 'hard water' when it passes over rocks containing calcium carbonate or magnesium carbonate.

Example:

calcium + carbonic → calcium hydrogen
carbonate   acid   carbonate
$$CaCO_3(s) + H_2CO_3(aq) \rightarrow Ca(HCO_3)_2(aq)$$

Calcium hydrogencarbonate solution is formed as the carbonate rock is dissolved away. This reaction helps to explain why caves are often found in limestone rocks.

If the calcium hydrogencarbonate solution is allowed to evaporate, calcium carbonate is again formed.

Example:

calcium hydrogen→ calcium + carbon + water
    carbonate   carbonate   dioxide
$$Ca(HCO_3)_2(aq) \rightarrow CaCO_3(s) + CO_2(g) + H_2O(l)$$

Water can be 'softened' by a number of methods, each remove Ca^{2+} or Mg^{2+} ions from solution.

1. *Boiling*—hydrogencarbonates of calcium and magnesium decompose, forming the solid carbonate. If the water can be softened by this method, it shows *temporary hardness*.

If the dissolved compounds are sulphates of magnesium or calcium, the hardness is '*permanent*'. Boiling will not remove it.

2. *Adding sodium carbonate*—this provides aqueous carbonate ions which then precipitate the metal carbonate.

Example:
$$Ca^{2+}(aq) + CO_3^{2-}(aq) \rightarrow CaCO_3(s)$$

3. *Ion exchange resins*—these are small beads which exchange ions. They remove Ca^{2+} and Mg^{2+} and replace them with H^+. Negative ions are exchanged, using another resin, for example removing SO_4^{2-} or HCO_3^- and adding OH^-.

17.6 Carbon monoxide

Carbon dioxide is a relatively harmless, even friendly gas. You do not need to take much care when you open a bottle of lemonade. You do not need to worry if you breathe in some of the carbon dioxide that escapes.

Carbon monoxide is far less friendly. It is poisonous and even breathing a very small amount can make you ill (see Section 13.3).

Can you suggest a property of carbon monoxide which makes it more dangerous than other poisonous gases?

Carbon monoxide is one of the main polluting gases as it is formed whenever any fuel burns in a limited supply of air. It is present in exhaust fumes and can be formed by oil or gas heaters unless the air supply is correctly adjusted. You may have heard people say "never reverse a car into a garage and leave the engine running." You should now understand why this is said.

Carbon monoxide is formed when steam is passed over white hot coke.

carbon + water → hydrogen + carbon
         (steam)   monoxide
$$C(s) + H_2O(g) \rightarrow H_2(g) + CO(g)$$

The mixture of hydrogen and carbon monoxide is known as water gas. It is used as a fuel because both gases burn—the products being carbon dioxide and steam.

Because carbon monoxide has a strong tendency to form carbon dioxide it is a good *reducing agent*.

The oxides of iron, lead and copper can be changed into the metal by heating in an atmosphere of carbon monoxide. This reaction can be demonstrated in a fume cupboard using the apparatus shown in Fig 17.7.

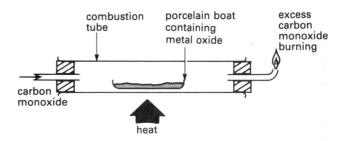

Fig 17.7 Reducing metal oxides with carbon monoxide

Example:

lead(II) + carbon → lead + carbon
 oxide   monoxide   dioxide
$$PbO + CO \rightarrow Pb + CO_2$$

You may remember that in the blast furnace iron ore is changed into iron by reacting with carbon monoxide.

At the start of this chapter we saw that diamond and graphite, both forms of carbon, are very different. Now we have seen that the two oxides of carbon are also very different. You might try to make a table showing four differences between carbon monoxide and carbon dioxide. Then try to find four ways in which they are similar.

Questions

1. When sodium hydroxide is left on a watch glass for several days, a white powder is formed. A colourless gas is given off when dilute hydrochloric acid is added to the powder. The powder is formed by the reaction between sodium hydroxide and:

A oxygen in the air
B water vapour in the air
C carbon dioxide and oxygen in the air
D water vapour and carbon dioxide in the air
E oxygen in the air.

2. Which one of the following gases burns in air?

A ammonia
B carbon monoxide
C hydrogen chloride
D oxygen
E sulphur dioxide

3. What is the volume of carbon dioxide formed when $20\,cm^3$ of methane (CH_4) burns completely in excess oxygen? (All volumes are measured at the same temperature and pressure.)

A $10\,cm^3$
B $20\,cm^3$
C $30\,cm^3$
D $40\,cm^3$
E $70\,cm^3$

4. Which one of the following processes does not produce carbon dioxide?

A burning of fossil fuels
B manufacture of iron
C manufacture of lime
D photosynthesis
E fermentation

5. When carbon dioxide is passed through an aqueous solution R, a white precipitate S is formed. When more carbon dioxide is passed through, the white precipitate disappears and a colourless solution T is formed. Which one of the following sets is R, S and T?

	R	S	T
A	$CaCO_3$	$Ca(HCO_3)_2$	$CaCO_3$
B	$Ca(HCO_3)_2$	$CaCO_3$	$Ca(OH)_2$
C	$Ca(HCO_3)_2$	$Ca(OH)_2$	$CaCO_3$
D	$Ca(OH)_2$	$CaCO_3$	$Ca(HCO_3)_2$
E	$Ca(OH)_2$	$Ca(HCO_3)_2$	$CaCO_3$

6. You have a supply of gas which contains hydrogen with a carbon dioxide impurity.

a How would you show that the gas contains carbon dioxide?
b What would you expect to see if the gas was bubbled through some distilled water containing Universal indicator?
c How could you obtain a pure sample of hydrogen from this gas supply?

7. Diamond and graphite are the allotropes of carbon.

a What are allotropes?
b Give two uses of diamond. What property of diamond makes it suitable for each of these uses?
c Give two uses of graphite. What property of graphite makes it suitable for each of these uses?
d How would you show that diamond and graphite are both pure forms of carbon? (Assume expense is no problem in this experiment.)

8.
a What property of carbon monoxide makes it poisonous to man?
b What gas is formed when carbon monoxide burns in air? Write an equation for the reaction.
c Why does city air usually contain a lower concentration of carbon monoxide on Sundays?
d One suggestion for reducing the carbon monoxide content of car exhaust fumes is to fit an extra section onto the exhaust pipe containing copper(II) oxide.

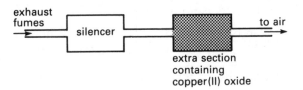

(i) How would this extra section reduce the carbon monoxide content of exhaust fumes?
(ii) Why do you think this method is not used? Give as many reasons as you can.

9. Carbon monoxide may be prepared in the laboratory by passing carbon dioxide over strongly heated carbon.

a Draw a labelled diagram of an apparatus which would be suitable for this purpose. You *must* show in your diagram how you would remove any unchanged carbon dioxide and how you would collect your carbon monoxide. You need *not* show the carbon dioxide generator.
b (i) Write an equation to illustrate the reaction taking place between the carbon dioxide and carbon.
(ii) Assuming that all the carbon dioxide used in the experiment is converted into carbon monoxide, what volume of carbon monoxide would be obtained from $100\,cm^3$ of carbon dioxide, under the same conditions of temperature and pressure?
c What would you expect to *see* if:
(i) a lighted taper was applied to a gas jar of carbon monoxide,

(ii) the resulting gas was shaken with lime water?

d Why is it dangerous to run a motor car engine in a closed garage?

[L]

10. Ethanedioic acid has the molecular formula $H_2C_2O_4$. The sodium salt, $Na_2C_2O_4$, is an ionic solid which is soluble in water. The barium salt, BaC_2O_4, is not soluble in water but forms a solution with dilute hydrochloric acid. When solid ethanedioic acid is warmed with concentrated sulphuric acid, a mixture containing equal volumes of carbon monoxide and carbon dioxide is given off.

a Draw a labelled diagram of the apparatus you would use to prepare and collect a sample of carbon monoxide free from carbon dioxide, starting from ethanedioic acid.

b Explain why equal volumes of carbon monoxide and carbon dioxide are formed when ethanedioic acid reacts with concentrated sulphuric acid. State clearly the function of the sulphuric acid in this reaction.

c Briefly explain the poisonous nature of carbon monoxide.

d Suggest reactions by which you could distinguish between aqueous solutions of sodium carbonate, sodium sulphate and the sodium salt of ethanedioic acid.

(ethanedioic acid = oxalic acid)

[C]

11.

a Name *two* crystalline forms of carbon.

b Describe experiments by which you could prepare good samples of *two* crystalline forms of sulphur from finely powdered sulphur.

c Weighed samples of carbon were heated in oxygen until combustion was complete. The carbon dioxide formed was absorbed in weighed bulbs containing potassium hydroxide solution. The experimental results were

	Mass of sample	Mass of CO_2 formed
I	0·40 g	1·47 g
II	0·60 g	1·76 g

(i) Give the equation for one reaction between carbon dioxide and potassium hydroxide.

(ii) Explain why potassium hydroxide solution was used rather than lime-water (calcium hydroxide solution).

(iii) Which sample, I or II, was pure carbon? Explain your reasoning.

[C]

12. Make use of the following information about silicon (Si) and its compounds to answer the questions below.

Silicon, atomic number 14, is the element immediately below carbon in Group IV of the Periodic Table. It does not react with water nor with dilute acids. It can be obtained by heating sand with an excess of magnesium. Sand is an oxide of silicon.

When sand is heated with carbon at a high temperature, carbon monoxide and carborundum are formed. Carborundum is a compound of silicon and carbon only and it is a very hard substance.

a State the characteristic valency of silicon and hence write down the chemical formulae for (i) sand, (ii) sodium silicate.

b Write the equation for the reaction between sand and magnesium.

c Describe how you would obtain pure *dry* silicon from the products of reaction **b**.

d How are the electrons arranged in the silicon atom?

e Give: (i) *two* physical differences, (ii) *one* chemical similarity, between carbon dioxide and sand.

f Suggest a formula for carborundum.

g Write the equation for the reaction of carbon with sand.

h Name another substance that you would expect to have the same crystal structure as carborundum.

[C]

18 Organic chemistry

Think of fuels like methane, petrol and diesel oil. Think of plastics like nylon, polythene and perspex. Think of medicines like aspirin and penicillin. Think of the protein, fat and carbohydrate that we need for a healthy diet. All these substances have one thing in common. They are all made up of *covalent, carbon-containing compounds*. These compounds are known as *organic chemicals*. The study of these compounds is known as *organic chemistry*.

There are many organic chemicals because carbon has the ability to form long chains of atoms.

18.1 Where do organic chemicals come from?

A number of organic chemicals are made from natural gas but the most important source of organic chemicals is oil. Oil obtained from underground deposits is known as *crude oil* (petroleum). It is a mixture of many different organic compounds. Crude oil itself is of little use, but the chemicals it contains are extremely useful.

It is the job of oil refineries to separate crude oil into its components. The first stage in this separation is to fractionally distil the crude oil.

The crude oil is completely vaporised by a furnace and the vapour passes into a fractionating column. The fractionating column is a tall tower made up of a number of compartments. These get cooler going up the column. As the vapour rises up the column, different substances condense in different compartments and the liquids formed are drawn off. The lower the boiling point of a substance the further it will travel up the column before condensing.

Generally speaking, the longer the chain of carbon atoms per molecule, the higher the boiling point of a substance.

The fractional distillation of crude oil does not produce pure substances, but it does provide a number of fractions. Each fraction contains a large number of chemicals that boil at a similar temperature. Some of these fractions can be used directly, but usually each fraction is further distilled and treated by various chemical processes to produce purer organic chemicals. One process is called 'cracking'. Fractions containing molecules with longer chains of carbon atoms are not as useful as those with shorter chains. In the 'cracking' process, the longer chain fraction is heated to a high temperature to break up the molecules into smaller pieces. The gas ethene (Section 18.4) is produced by this method.

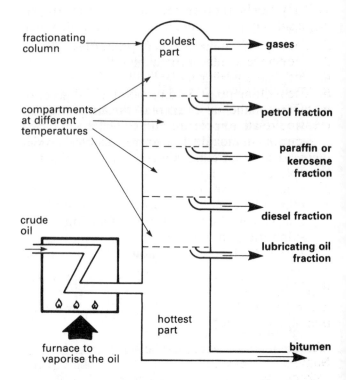

Fig 18.1 Fractional distillation of crude oil

Table 1 shows some of the uses of the fractions of crude oil.

Name of fraction	Use
Gases	Used as fuels such as bottled gas and lighter fuel
Petrol fraction	Used to make petrol (gasoline) for cars
Kerosene (Paraffin) fraction	Used for oil stoves and aircraft fuel
Diesel fraction	Used as a fuel for diesel engines
Lubricating oil fraction	Used as a lubricant and to make waxes and polishes
Bitumen	Used for road making and for sealing roofs

Table 1

18.2 Homologous series

Organic chemicals have a number of things in common:

They all contain carbon; most contain hydrogen.
They are nearly all covalent compounds.
They nearly all burn or char when heated in air.

However, organic chemicals can be as different as TNT and candlewax. Because so many organic chemicals exist, we divide them up into groups or families of compounds with similar properties. These families are known as *homologous series*.

Compounds which are members of the same homologous series have a number of things in common:

1. They can be represented by a general formula.
Example: C_nH_{2n}.
2. They have similar chemical properties.
3. They have similar structures.
4. They have similar names.
5. The melting point, boiling point and density of members of the series increase steadily with increasing relative molecular mass.

We can understand the properties of a homologous series more clearly if we consider a few examples.

18.3 Alkanes

The alkanes are *hydrocarbons*. This means that they are compounds that contain carbon and hydrogen only. Table 2 provides some information about the alkanes.

Name of alkane	Formula	Relative molecular mass	Boiling point /°C
methane	CH_4	16	−164
ethane	C_2H_6	30	−87
propane	C_3H_8	44	−42
butane	C_4H_{10}	58	0
pentane	C_5H_{12}	72	36
hexane	C_6H_{14}	86	69

Table 2 The alkanes

You should notice from Table 2 that:

1. All the alkanes have similar names. They all end in -ane.
The start of each name is different for each alkane. The start of the name tells us how many carbon atoms there are in one molecule of the alkane; meth means 1, eth means 2, prop means 3 and so on.
2. The alkanes can be represented by the general formula C_nH_{2n+2}, where n is the number of carbon atoms in a molecule of the alkane.
When $n = 1$, formula $= C_1H_{2+2} = CH_4$.
When $n = 4$, formula $= C_4H_{8+2} = C_4H_{10}$.
3. As the relative molecular mass increases the boiling point of the alkanes also increases.

We can see more similarities between the members of the alkane series if we consider the structures of the various compounds.

Name of Alkane	Molecular formula	Full structural formula
methane	CH_4	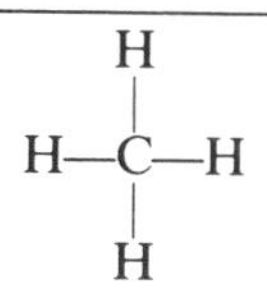
ethane	C_2H_6	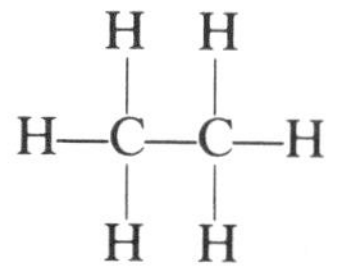
propane	C_3H_8	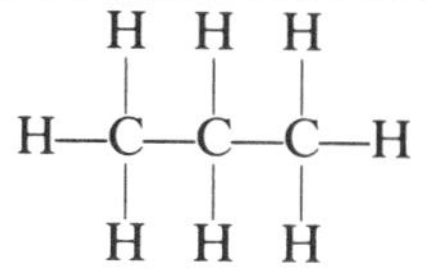
butane	C_4H_{10}	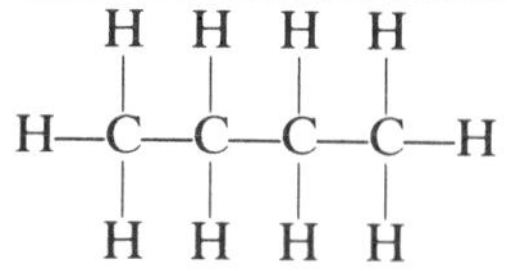 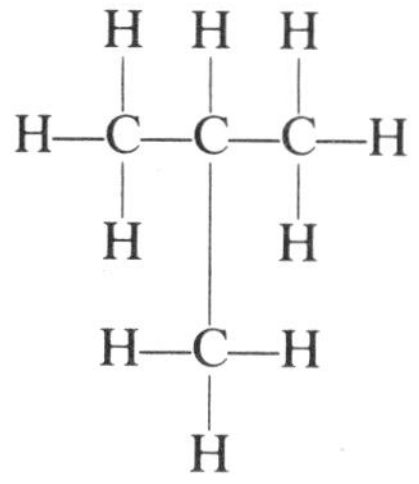
pentane	C_5H_{12}	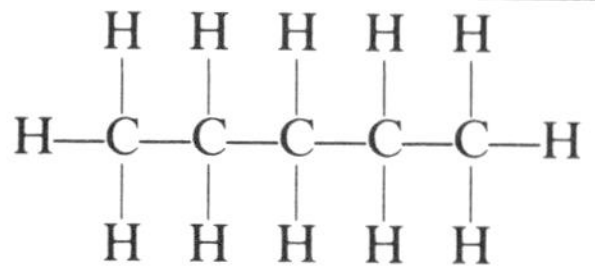 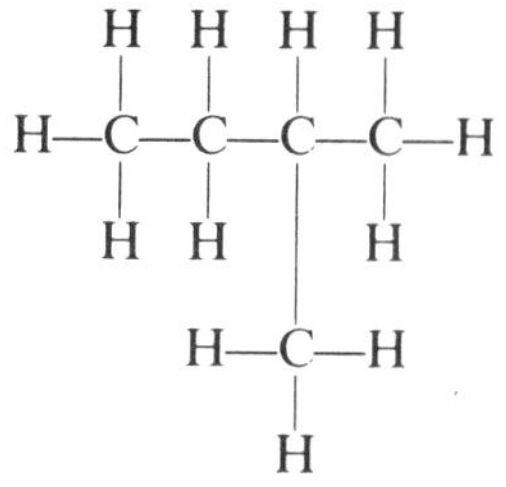 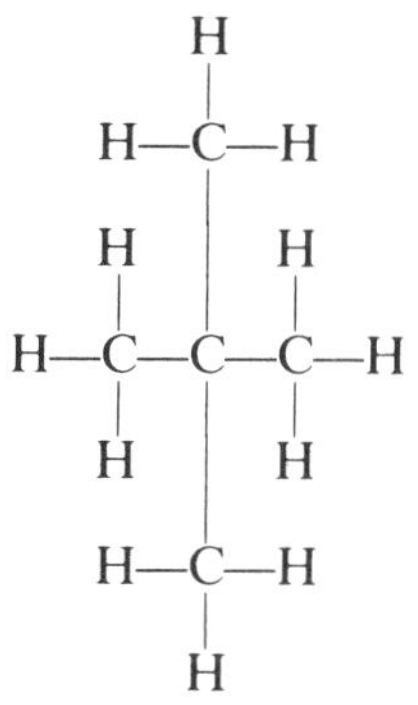

Table 3 Structure of the alkanes

Note The *molecular formula* shows how many atoms of each element there are in a molecule. *Example:* 1 molecule of methane contains 1 carbon atom and 4 hydrogen atoms. It therefore has a molecular formula CH_4.

The *full structural formula* shows how the atoms are arranged in the molecule.

Looking at Table 3 you should notice that:

1. The alkanes have similar structures. Each alkane (except methane) has a single bond between carbon atoms. Different homologous series have other similarities in their structures.

2. There are two possible structural formulae for butane. One has the carbon atoms arranged in a row C—C—C—C, whereas the other has a branched arrangement of carbon atoms

$$C—C—C$$
$$|$$
$$C$$

These different forms of butane are known as the *isomers* of butane. You can see that pentane has three isomers. *Isomers are compounds with the same molecular formula but different full structural formulae.*

It is important to realise that

$$\begin{array}{cccc} H & H & H & H \\ | & | & | & | \\ H—C—C—C—C—H \\ | & | & | & | \\ H & H & H & H \end{array}$$

and

$$\begin{array}{c} H \\ | \\ H\ H—C—H \\ | \\ H—C\quad\ C—H \\ | \\ H—C—H\ H \\ | \\ H \end{array}$$

are *not* different isomers—one is just a twisted form of the other.

Chemical properties of the alkanes

The alkanes are not a very reactive family of compounds. They do not take part in many chemical reactions, but they all react similarly when they do react.

1. All alkanes will burn in a plentiful supply of air to form carbon dioxide and water.

Example:

methane + oxygen → carbon dioxide + water
$$CH_4\ +\ 2O_2\ \rightarrow\ CO_2\ +\ 2H_2O$$

2. They react with chlorine or bromine in the presence of sunlight.

Example:

methane + chlorine $\xrightarrow{\text{sunlight}}$ chloro- + hydrogen
methane chloride
$$CH_4\ +\ Cl_2\ \longrightarrow\ CH_3Cl\ +\ HCl$$

This is a *substitution* reaction in which one hydrogen atom in a methane molecule is substituted by a chlorine atom.

Further substitution reactions can occur so that eventually all the hydrogen atoms have been replaced to form tetrachloromethane, CCl_4.

3. The alkanes *do not* decolorise potassium manganate(VII) solution and do not normally decolorise bromine water.

Uses of the alkanes

1. They are mainly used as fuels:
Methane is the main component of natural gas.
Butane is the main component of camping gas and lighter fuel.
Octane is a component of petrol.
2. They are used to produce alkenes, such as ethene by a process called cracking.

18.4 Alkenes—a more reactive family

The alkenes are a homologous series with the general formula C_nH_{2n}. All alkenes contain a *double bond* between two carbon atoms. This carbon-carbon double bond makes the alkenes far more reactive than the alkanes.

By looking at Table 4 you should notice that:

1. All alkenes have names ending in -ene.
2. All alkenes contain carbon-carbon double bonds.
3. Butene has three isomers and pentene has four isomers. This is because the double bond can have different positions along the carbon chain and also the chain can be branched.

You will have noticed that the names of isomers of C_4H_8 and C_5H_{10} have numbers in them. These numbers tell us where the double bond is. The carbon atoms are numbered from one end of the molecule:

$$^1C—^2C—^3C—^4C$$

The name but-1-ene (read as 'but-one-ene') means that the double bond is between carbon atoms 1 and 2. The name but-2-ene means that the double bond is between carbon atoms 2 and 3.

Chemical properties of alkenes

All alkenes burn in a plentiful supply of air to form carbon dioxide and water.

Example:

ethene + oxygen → carbon dioxide + water
$$C_2H_4\ +\ 3O_2\ \rightarrow\ 2CO_2\ +\ 2H_2O$$

When the alkenes burn they release a large amount of energy. This would make them good fuels, but the alkenes have chemical properties that make them far too useful to burn.

Addition reactions of alkenes

In the presence of a catalyst (such as nickel), alkenes react with hydrogen to form alkanes:

ethene + hydrogen → ethane

$$\begin{array}{c} H \qquad H \\ \diagdown \qquad \diagup \\ C＝C\ +\ H_2\ \xrightarrow{\text{Ni}}\ \begin{array}{cc} H & H \\ | & | \\ H—C—C—H \\ | & | \\ H & H \end{array} \\ \diagup \qquad \diagdown \\ H \qquad H \end{array}$$

Name of alkene	Molecular formula	Full structural formula
ethene	C_2H_4	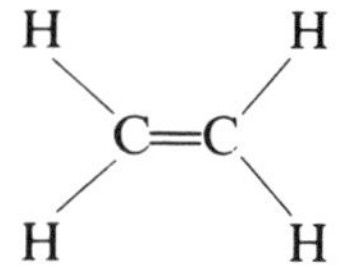
propene	C_3H_6	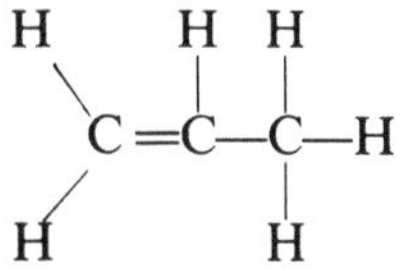
butene	C_4H_8	but-1-ene 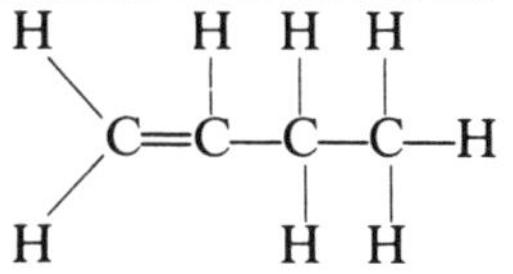 *or* but-2-ene 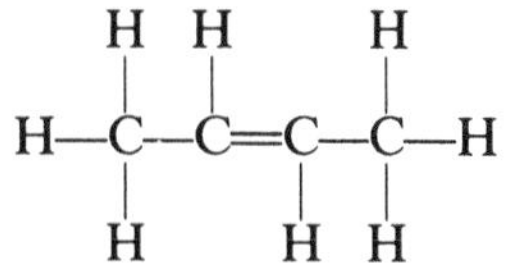 *or* 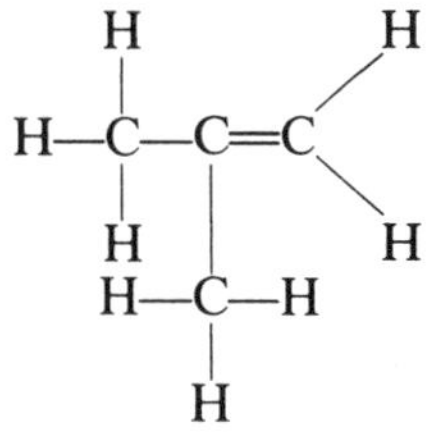
pentene	C_5H_{10}	pent-1-ene 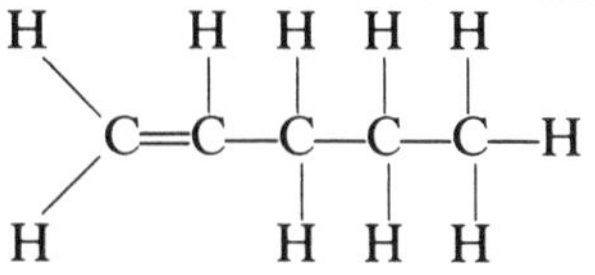 *or* pent-2-ene 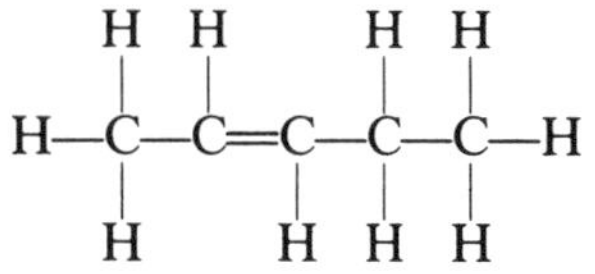 *or* 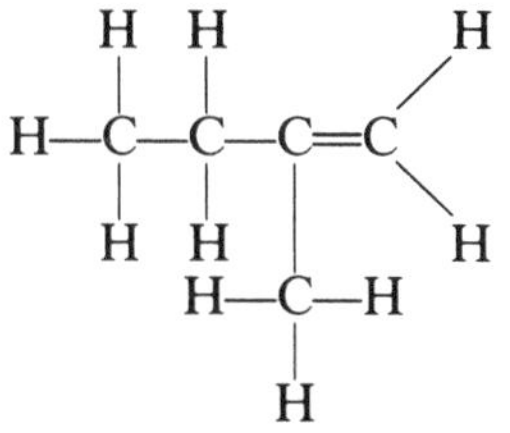 *or* 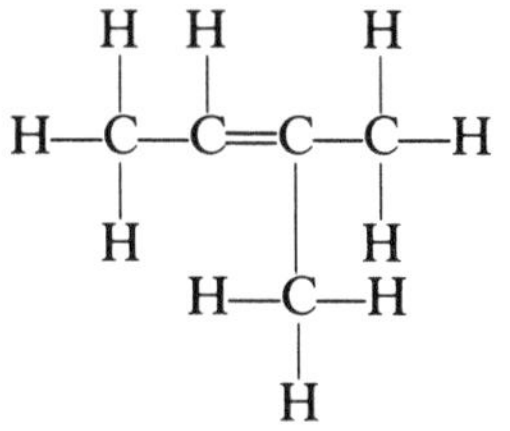

Table 4 The alkenes

The addition of hydrogen to carbon–carbon double bonds is an important stage in changing vegetable oils into margarine.

Alkenes react with bromine or bromine water very rapidly. The bromine is decolorised and a colourless oil is formed:

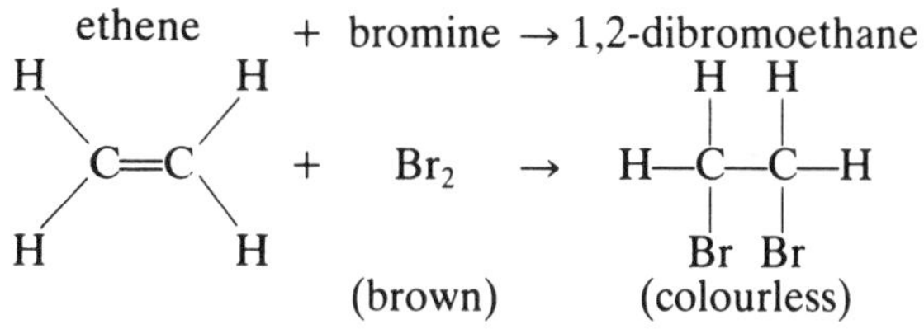
ethene + bromine → 1,2-dibromoethane
+ Br2 →
(brown)
(colourless)

(A similar reaction occurs with chlorine.)

Ethene will react with steam when passed over a catalyst of phosphoric(V) acid on silica at 300°C and a pressure of 60 atmospheres.

ethene + steam → ethanol

$$\begin{array}{ccc} \text{H} & & \text{H} \\ & \diagdown & \\ & \text{C}=\text{C} & \\ & \diagup & \\ \text{H} & & \text{H} \end{array} + \text{H}_2\text{O} \rightarrow \text{H}-\text{C}-\text{C}-\text{O}-\text{H}$$

It can be seen from the equations that the attacking molecules add onto the alkene molecule to form a larger molecule. The carbon-carbon double bond allows the alkene to react by *addition*. Chemists say that alkenes are *unsaturated* substances because they can react by addition.

All alkenes will decolorise bromine. This is the usual test for an unsaturated compound.

Another useful test involves acidified potassium manganate(VII). This turns from purple to colourless.

Alkenes are a very important group of compounds. They are the starting materials for an enormous number of organic compounds, including plastics and solvents.

Experiment 18.1 Comparing alkanes and alkenes

You will need: a sample of an alkane and an alkene (hexane and cyclohexene are suitable), a dilute solution of bromine in 1,1,1-trichloroethane (a solvent) and a very dilute solution of potassium manganate(VII) acidified with dilute sulphuric acid.

1. To separate small samples of each hydrocarbon in a test tube add about $2\,\text{cm}^3$ of water. Shake the mixture and then allow it to stand.

Do the hydrocarbons dissolve in water? Are these organic compounds less dense than water?

2. In separate test tubes add a few drops of each to about $2\,\text{cm}^3$ of the solution of bromine. Shake each tube.

Which hydrocarbon reacts more rapidly?

3. Repeat test 2 but using the potassium manganate(VII) solution instead of the bromine solution.

Which hydrocarbon reacts more rapidly?

18.5 Alcohols

The alcohols are a homologous series with the general formula $C_nH_{2n+1}OH$. All alcohols contain an oxygen atom joined on one side to a carbon atom and on the other to a hydrogen atom. This —O—H group (*hydroxyl group*) makes the alcohols fairly reactive.

In Table 5, you will see that all alcohols have names ending in -ol. Notice also, that the beginning of the name tells us the number of carbon atoms in a molecule of the alcohol.

Name of alcohol	Molecular formula	Full structural formula
methanol	CH_3OH	$\begin{array}{c} \text{H} \\ \mid \\ \text{H}-\text{C}-\text{O}-\text{H} \\ \mid \\ \text{H} \end{array}$
ethanol	C_2H_5OH	$\begin{array}{c} \text{H}\ \ \text{H} \\ \mid\ \ \mid \\ \text{H}-\text{C}-\text{C}-\text{O}-\text{H} \\ \mid\ \ \mid \\ \text{H}\ \ \text{H} \end{array}$
propanol	C_3H_7OH	$\begin{array}{c} \text{H}\ \ \text{H}\ \ \text{H} \\ \mid\ \ \mid\ \ \mid \\ \text{H}-\text{C}-\text{C}-\text{C}-\text{O}-\text{H} \\ \mid\ \ \mid\ \ \mid \\ \text{H}\ \ \text{H}\ \ \text{H} \end{array}$
butanol	C_4H_9OH	$\begin{array}{c} \text{H}\ \ \text{H}\ \ \text{H}\ \ \text{H} \\ \mid\ \ \mid\ \ \mid\ \ \mid \\ \text{H}-\text{C}-\text{C}-\text{C}-\text{C}-\text{O}-\text{H} \\ \mid\ \ \mid\ \ \mid\ \ \mid \\ \text{H}\ \ \text{H}\ \ \text{H}\ \ \text{H} \end{array}$
pentanol	$C_5H_{11}OH$	$\begin{array}{c} \text{H}\ \ \text{H}\ \ \text{H}\ \ \text{H}\ \ \text{H} \\ \mid\ \ \mid\ \ \mid\ \ \mid\ \ \mid \\ \text{H}-\text{C}-\text{C}-\text{C}-\text{C}-\text{C}-\text{O}-\text{H} \\ \mid\ \ \mid\ \ \mid\ \ \mid\ \ \mid \\ \text{H}\ \ \text{H}\ \ \text{H}\ \ \text{H}\ \ \text{H} \end{array}$

Table 5 (Propanol, butanol and pentanol have ⁀her isomers.)

Chemical properties of alcohols

All alcohols burn in a plentiful supply of air to form carbon dioxide and water.

Example:

ethanol + oxygen → carbon dioxide + water

$$C_2H_5OH + 3O_2 \rightarrow 2CO_2 + 3H_2O$$

When alcohols burn they release a large amount of energy. In some countries, cars have been converted to run on ethanol instead of petrol.

Ethanol can be oxidised to ethanoic acid (the main ingredient of vinegar). Acidified potassium dichromate(VI) is a suitable oxidising agent. It turns from orange to green during the reaction.

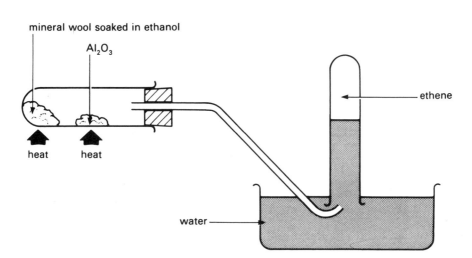

[O] represents oxygen from an oxidising agent, it is not meant to be an oxygen atom. Full equations involving potassium dichromate(VI) are complicated. The use of [O] makes writing equations easier.

Ethanol is slowly oxidised by air to ethanoic acid. That is why drinks containing ethanol (such as wine and beer), eventually turn sour if they are not stored correctly.

Sodium reacts with alcohols to produce hydrogen. This is because this type of compound contains the —O—H group.

Water contains the same arrangement of atoms (H—O—H), so sodium will react with water and alcohols, to give hydrogen. The reaction with alcohols is much less vigorous than that with water.

The sodium ethoxide is obtained as a white solid if excess ethanol is evaporated. The fact that it is a solid suggests that the bonding between the sodium and the ethoxide group (C_2H_5O), is ionic.

Other reagents that react with —O—H groups are the chlorides of phosphorus (PCl_3 and PCl_5). These replace the hydroxyl group with a chlorine atom, hydrogen chloride gas being produced.

Example:

$$C_2H_5OH + PCl_5 \rightarrow C_2H_5Cl + HCl + POCl_3$$
ethanol   chloroethane

Ethanol is formed by an addition reaction of ethene with steam.

The opposite of this reaction, *dehydration*, is done by heating with concentrated sulphuric acid. It is more convenient, however, to carry out this reaction using aluminium oxide (as catalyst) in the apparatus shown in Fig 18.2.

$$C_2H_5OH \rightarrow C_2H_4 + H_2O$$

You can collect several test tubes of gas by this method. Test with bromine water to show that an alkene is present.

18.6 Carboxylic acids

The carboxylic acids are a homologous series of compounds with the general formula $C_nH_{2n+1}CO_2H$.

Chemical properties of carboxylic acids

Because all carboxylic acids contain the —O—H group, they all react with sodium to produce hydrogen.

Carboxylic acids are usually weak acids, but they show the reactions of a typical acid.

Fig 18.2

Name of acid	Molecular formula	Full structural formula
methanoic acid	HCO_2H	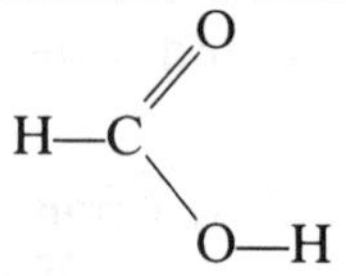
ethanoic acid	CH_3CO_2H	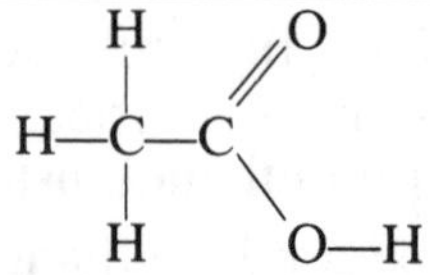
propanoic acid	$C_2H_5CO_2H$	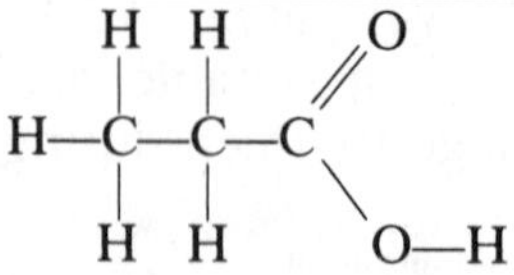
butanoic acid	$C_3H_7CO_2H$	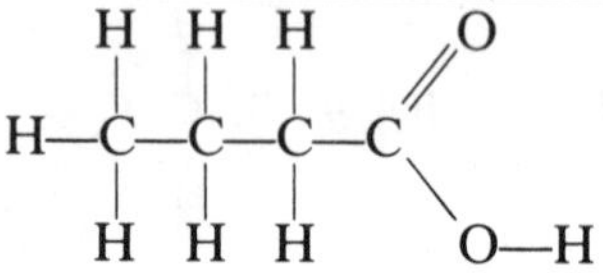
pentanoic acid	$C_4H_9CO_2H$	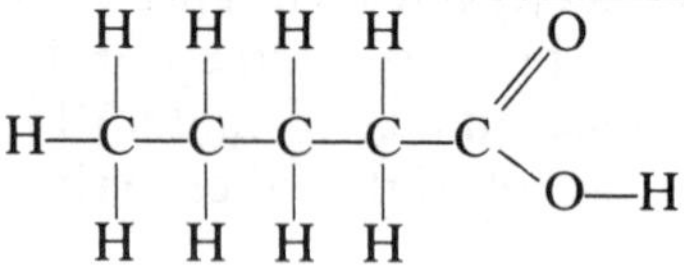

Table 6 The beginning of the name of each acid tells how many carbon atoms there are in each molecule.

They react with metals to give the metal salt and hydrogen (as shown by the reaction with sodium).
They react with metal oxides and hydroxides to give the salt and water.
They react with carbonates to give the salt, carbon dioxide and water.

These organic acids also react with alcohols to form *esters* and water.

ORGANIC ACID + ALCOHOL ⇌ ESTER + WATER

Example:

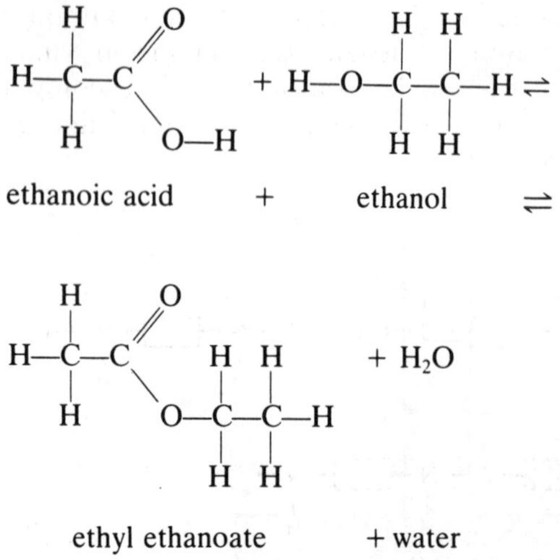

ethanoic acid   +   ethanol   ⇌

ethyl ethanoate   + water

The reaction is, in fact, reversible and a few drops of concentrated sulphuric acid are usually added as a catalyst. After warming for a while, a sweet fruity smell is detected. This smell is common with esters. Some of them are used as food flavourings and others are used in perfumes.

In naming esters, the first word tells us the group attached to the oxygen atom. The second word, tells us what acid has been used.

Ethyl ethanoate means that there is an ethyl group (C_2H_5) attached to the oxygen atom. The word ethanoate tells us that ethanoic acid has been used.

Methanol and butanoic acid react to give methyl butanoate.

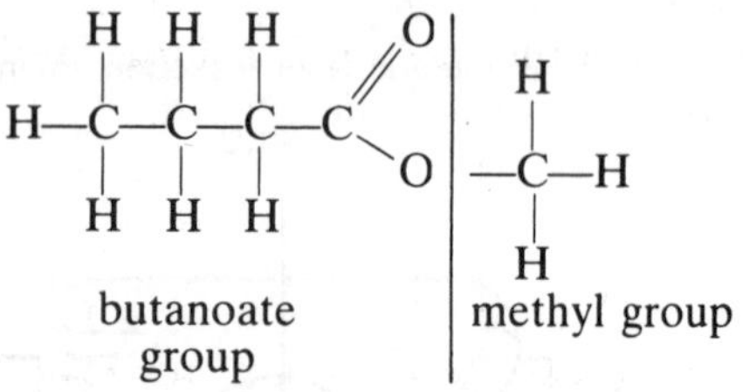

butanoate   methyl group
group

Experiment 18.2 Reactions of ethanol and ethanoic acid

You will need: ethanol, aqueous ethanoic acid, magnesium ribbon, chips of marble. You will also need dilute hydrochloric acid and aqueous potassium dichromate(VI) acidified with dilute sulphuric acid.

1. Add a few drops of ethanol to $5\,cm^3$ of the aqueous potassium dichromate(VI). Place the test tube in a beaker of hot water for 10 minutes.

What do you observe and what organic substance is produced in the reaction?

2. In separate test tubes put about $5\,cm^3$ of aqueous ethanoic acid and $5\,cm^3$ dilute hydrochloric acid. To each add a small piece of magnesium ribbon.

What gas is produced?

3. Repeat test 2 but use marble chips instead of the magnesium ribbon.

What gas is produced?

How can you explain the different rates of reaction shown by ethanoic acid and dilute hydrochloric acid in test **2** and **3**?

18.7 Macromolecules

These are large molecules that can be considered to be made up from small units.

1. *Synthetic Polymers*

We have seen that the alkenes can react by addition. What is really important is that they can 'add on' to one another, time and time again to form very long molecules. The large molecule formed when we start from ethene is a poly(ethene) molecule or polythene (the plastic).

For ethene to form poly(ethene), one bond in each ethene molecule must break, and new bonds must be formed, to hold the ethene molecules together. The arrows in the following equation show how this happens.

ethene $\longrightarrow$ poly(ethene)

$$CH_2{=}CH_2 + CH_2{=}CH_2 + CH_2{=}CH_2 \rightarrow$$
$$-CH_2-CH_2-CH_2-CH_2-CH_2-CH_2-$$

The plastic polythene is made up of very large *poly(ethene)* molecules. Each poly(ethene) molecule is made up of thousands of ethene molecules joined together.

The process in which many small molecules add on to one another to form a large molecule, is known as *addition polymerisation*.

The starting small molecule is known as the *monomer*. The final large molecule is known as the *polymer*.

Any alkene should be capable of acting as a monomer for addition polymerisation. By varying the monomer, different polymers or plastics are formed with different properties.

We have seen that one method of making plastics is by a process known as addition polymerisation. Different plastics can be made by a process known as *condensation polymerisation*.

Condensation polymerisation is different from addition polymerisation in a number of ways:

1. In condensation polymerisation there are usually two different monomers reacting together.
2. In condensation polymerisation a small molecule (usually water) is thrown out every time two monomer molecules react.

Two examples of plastics formed by condensation polymerisation are *nylon* and *Terylene*.

The two monomers used to make nylon are represented by the diagrams below. The boxes in Fig 18.3 represent a unit of carbon atoms, e.g.

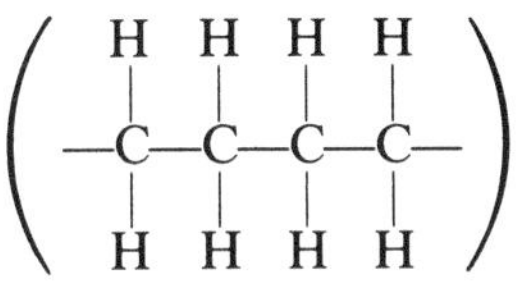

$$\left(\begin{array}{c} H\ H\ H\ H \\ |\ \ |\ \ |\ \ | \\ -C-C-C-C- \\ |\ \ |\ \ |\ \ | \\ H\ H\ H\ H \end{array} \right)$$

Every time two units join together a molecule of water is formed. The links

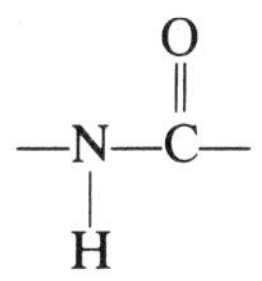

$$\begin{array}{c} O \\ \| \\ -N-C- \\ | \\ H \end{array}$$

between the units are called *amide linkages*, so nylon is called a *polyamide*. Nylon can be made into fibres and it is used to make ropes and clothing.

Experiment 18.3 The preparation of nylon

You will need: a 10% solution of decanedioyl dichloride in 1,1,1-trichloroethane (solution A), a 5% solution of hexane-1,6-diamine in $1\,mol/dm^3$ sodium hydroxide (solution B). The compounds used in this experiment are *harmful* and *gloves must be worn*.

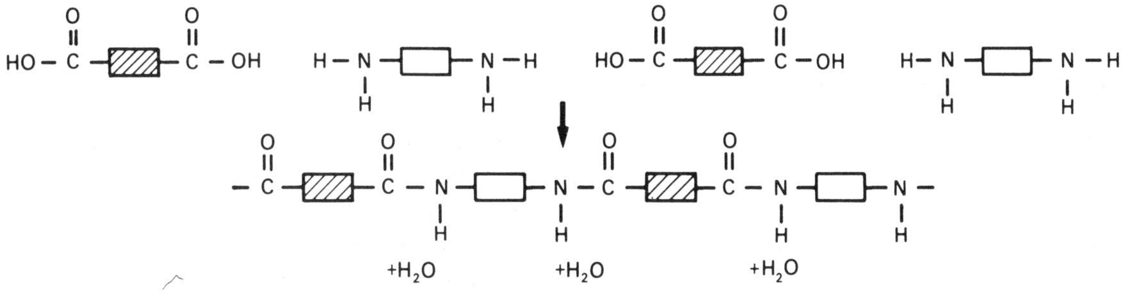

Fig 18.3

125

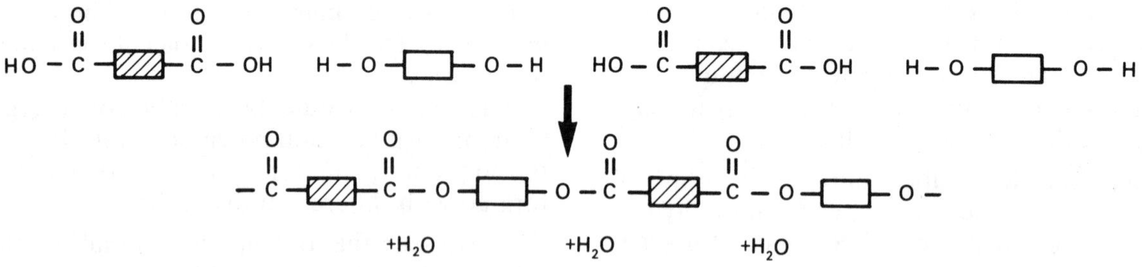

Fig 18.4

Pour $10\,cm^3$ of solution A into a small beaker. Carefully pour $10\,cm^3$ of solution B down the side of the beaker so that the solution floats on top of solution A.

Look where the two solutions meet. You should see a thin film of nylon. Using a glass rod, carefully lift the centre of the film and wind it around the rod. As you wind up the polymer more is formed as the two solutions come into contact.

What type of covalent linkage joins the carbon units together in nylon?

Is nylon a condensation polymer or an addition polymer?

The formation of *Terylene* from its two monomers is shown in Fig 18.4.

Terylene is made by joining acid units to alcohol units and a polymer is formed having ester linkages

$$-\overset{\overset{\displaystyle O}{\|}}{C}-O-$$

between the units. *Terylene* is called a *polyester*, and is used to make clothing.

2. Natural Macromolecules

Three important classes of compound in our diets are proteins, fats and carbohydrates. Although not all of these can be classed as polymers, they do have some similarities. They all have large molecules.

Proteins

These have the same amide linkages as nylon but the units are different. The structure of a protein, which is a polyamide, is shown below:

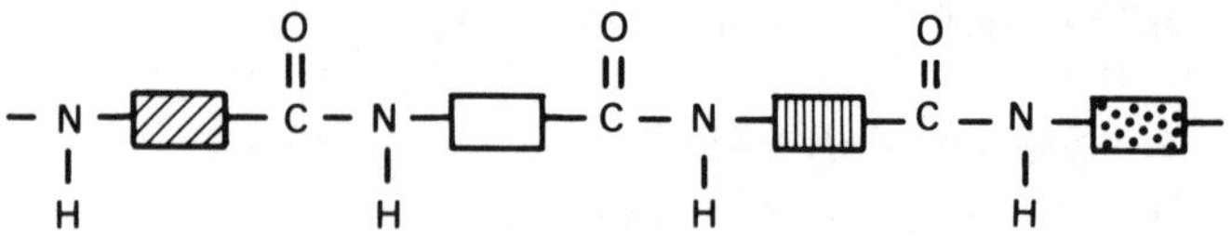

Fig 18.5

Proteins can be broken down into their component units (*aminoacids*), by heating with dilute acid or alkali.

This process is known as *hydrolysis* and is the opposite of condensation polymerisation (Fig 18.6).

The aminoacids obtained can be separated and identified by using paper chromatography (Section 2.1).

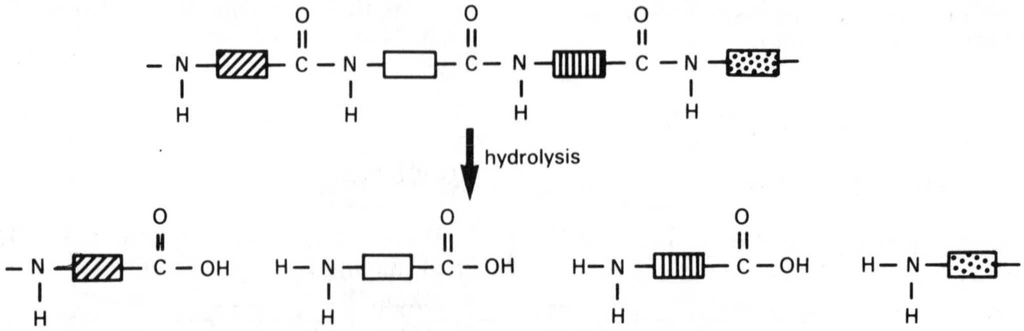

hydrolysis

Fig 18.6

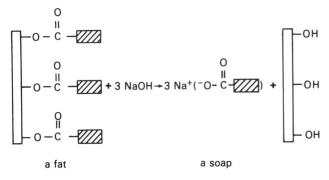

a fat   a soap

Fig 18.7

Fats

These have ester linkages, like *Terylene* but the units are not the same, as their arrangement is slightly different. A typical fat is shown above.

Compounds of this type (such as palm oil) can be hydrolysed with sodium hydroxide to produce soap.

Detergents are usually made from hydrocarbons using concentrated sulphuric acid.

Soaps have the unfortunate property that they form a scum with hard water (Section 17.5), whereas synthetic detergents do not.

As detergents are now used so widely, they can cause pollution problems. Some are not affected by the processes involved in sewage treatment. They can be pollutants in rivers, as they are toxic to fish. Soaps made from oils, such as palm oil, are broken down by bacteria in the treatment of sewage and do not cause pollution.

Carbohydrates

Carbohydrates are compounds that contain carbon, hydrogen and oxygen only. This class of compound is made up of a large number of sugar units joined together by oxygen atoms:

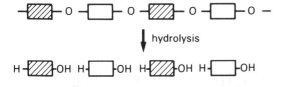

Carbohydrates, such as starch, can be broken down by heating with dilute acid. This process of hydrolysis produces a mixture of sugars:

Fig 18.8

These smaller molecules can be separated and identified using chromatography.

Sugars can be broken down still further, by using a process called *fermentation*. In this process, yeast is added to a sugar solution and the products are ethanol and carbon dioxide. The yeast contains biological catalysts known as enzymes.

sugar $\xrightarrow{\text{yeast}}$ ethanol + carbon dioxide

Example:

$$C_6H_{12}O_6 \xrightarrow{\text{yeast}} 2C_2H_5OH + 2CO_2$$

This type of process is used in making beer and wine.

Fermentation can be carried out using the apparatus shown.

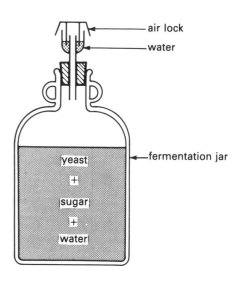

Fig 18.9 Fermenting sugar solution

The air above the solution is gradually replaced by the carbon dioxide produced. The air lock prevents air from entering the vessel—this could lead to the oxidation of ethanol to ethanoic acid.

An almost pure sample of ethanol can be obtained from the mixture by filtering off the yeast and then fractionally distilling the solution. Ethanol has a lower boiling point than water and therefore the liquid collected first is almost pure ethanol.

Questions

1. Which one of the following is the best method of distinguishing between an alkane and an alkene?

    A add bromine water
    B add lime-water
    C burn the gases
    D test their solubility in water
    E test with pH paper

127

2. Which one of the following reacts with sodium hydroxide to form a soap?

A   an alcohol
B   a carbohydrate
C   a fat
D   a hydrocarbon
E   a protein

3. Which one of the following compounds does not react with sodium to give hydrogen?

A H_2O
B CH_3OH
C CH_3CO_2H
D C_2H_5OH
E $CH_3CO_2C_2H_5$

4. The structure below represents the structure of *Terylene*.

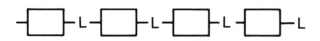

The linkage L is

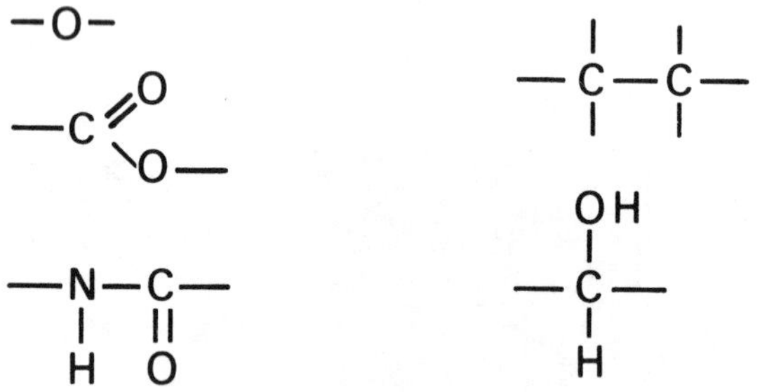

$-O-$

$-C \overset{\displaystyle O}{\underset{\displaystyle O-}{<}}$

$\underset{\displaystyle \overset{|}{H} \ \overset{|}{O}}{-N-C-}$

$\overset{\displaystyle |\ \ |}{\underset{\displaystyle |\ \ |}{-C-C-}}$

$\overset{\displaystyle O H}{\underset{\displaystyle \overset{|}{H}}{-C-}}$

5. A hydrocarbon burns completely in oxygen to form 8·8 g of carbon dioxide and 4·5 g of water. What is the formula of this hydrocarbon?

A C_4H_8
B C_4H_{10}
C C_5H_{10}
D C_5H_{12}
E C_6H_{12}

6. BUTANE, ETHANOL, ETHENE, METHANE, POLY(ETHENE), SUGAR.
Choosing from the above list, name:

a two hydrocarbons which are in the same homologous series
b a hydrocarbon that will react with bromine water
c a plastic
d an alkane
e an alkene
f an alcohol
g a carbohydrate
h a hydrocarbon with four carbon atoms per molecule
   a substance formed by fermentation.

7.
a What name is given to organic compounds which contain carbon and hydrogen only?
b Draw the full structural formulae of:
   (i) two compounds with the molecular formula C_4H_{10}
   (ii) three compounds with the molecular formula C_5H_{12}.
c What name is given to substances with the same molecular formula but different structural formulae?

8.
a Ethene is an unsaturated compound. Which chemical reaction can be used to show that ethene is unsaturated?
b Draw the full structural formula of ethene. What part of the structure shows that ethene is unsaturated?
c Ethene is the monomer from which the polymer poly(ethene) is made. Draw a diagram to show the structure of part of a poly(ethene) molecule.

9.
a Name three important substances obtained from crude oil and give a major use of each.
b Name the process by which crude oil is separated into its components.
c What is the name of the process used to break down saturated hydrocarbon molecules into smaller unsaturated molecules? Why is this process so important?
d What is the name of the process by which small unsaturated hydrocarbon molecules add on to each other to form long chain saturated molecules?

10. FATS, CARBOHYDRATES, PROTEIN.
Choosing from the above list, name:
a a type of food that burns to form carbon dioxide and water only
b a type of food that contains nitrogen
c a type of food that can be hydrated to produce sugars.
d a type of food which is digested to form amino acids
e a type of food that is made up of long chain molecules.

11. One of the hydrocarbons in petrol has the molecular formula C_8H_{18}. When petrol is burned in a motor car engine the exhaust fumes contain approximately 9% carbon dioxide, 5% carbon monoxide, 4% oxygen, 2% hydrogen, 0·2% hydrocarbons and 0·2% oxides of nitrogen.

a What is meant by the term "hydrocarbon"?
b The gases listed above make up 20·4% of the composition of the exhaust fumes. What gas makes up the greater percentage of the remaining gases?

c Suggest a reason for the presence of: (i) carbon dioxide, (ii) carbon monoxide, (iii) oxides of nitrogen, in the exhaust fumes.

d Butene and butane can be formed by a process known as cracking, whereby a larger hydrocarbon molecule is broken down by high temperature.

For example: $C_8H_{18}(g) \rightarrow C_4H_8(g) + C_4H_{10}(g)$

(i) Is C_8H_{18} an alkane or an alkene? Explain your answer. Give one chemical test by which you could distinguish between an alkene and an alkane.

(ii) Write an equation to show how hydrogen could be formed from C_4H_{10} by cracking.

e Give *two* reasons why car exhausts pipes tend to rust more quickly than the bodywork although both are made of steel.

f Assuming that air contains one-fifth of oxygen by volume, calculate the minimum volume of air at room temperature and pressure needed for the complete combustion of 57 g of the hydrocarbon of molecular formula C_8H_{18}.

[C]

12.

a (i) Name the essential material which must be added to an aqueous solution of sucrose in order to prepare an aqueous solution of ethanol. State the name given to this type of process.

(ii) Give a diagram of the apparatus you would use to obtain a more concentrated solution of ethanol from the solution produced in (i).

b Ethanol can be manufactured from ethene. For this reaction give the essential reaction conditions and write the equation.

c Ethyl ethanoate $(CH_3CO_2C_2H_5)$ can be prepared from ethanol and ethanoic acid.

(i) Write the equation for the reaction.

(ii) Calculate the theoretical yield of ethyl ethanoate which can be obtained from 23 g of ethanol.

(iii) Experimentally it was found that 33 g of ethyl ethanoate were obtained from 23 g of ethanol. Calculate the percentage yield.

d Ethanol is the second member of the homologous series of alcohols.

(i) Give the name and full structural formula of the first member of this series.

(ii) Give three general properties of an homologous series.

[C]

13. A, CH_3OH; **B,** CH_3CO_2H; **C,** $CH_3CH=CH_2$; **D,** CH_3CH_2OH; **E,** $CH_3CH_2CH_2CH_3$.

a For each of the compounds **A** to **E** in the list above, give its name, the name and general formula of the homologous series to which it belongs and state whether the compound is a solid, liquid or gas at room temperature and pressure.

b On complete combustion, 0·100 mol of one of the compounds **A** to **E** gave 13·2 g of carbon dioxide. Identify the compound, explaining your reasoning.

c What is meant by *an ester*? Using only compounds chosen from **A** to **E**, describe how you would prepare an ester. Write the equation for the preparation.

[C]

14.

a Soaps can be made by heating fats, e.g. glyceryl stearate, with aqueous sodium hydroxide.

(i) Complete the following equations in words:

glyceryl stearate + sodium hydroxide →

(ii) What name do chemists give to this type of reaction?

(iii) To which class of organic compound do fats, such as glyceryl stearate, belong?

(iv) Give the name of the substance which is added to make the soap less soluble in the resulting mixture.

b (i) Give the names of two chemicals which could be used in the laboratory to prepare a sample of a synthetic detergent.

(ii) Give the name of a raw material from which a detergent might be manufactured on an industrial scale.

(iii) Give *two* advantages which modern detergents have over soaps.

(iv) Why is it sometimes said that it is unwise to use detergents for dispersing oil slicks?

[L]

15.

a Each of the following conversions can be carried out in the laboratory:

Glucose   → Ethanol
Ethanol   → Ethanoic acid (Acetic acid)
Ethanoic Acid → Ethyl ethanoate
                      (Ethyl acetate)

Using simple test tube experiments, describe how you would carry out these conversions in the laboratory. In your account you should state clearly the starting materials you would use, the conditions necessary for the reaction to take place, any observations you would expect to make and how you would attempt to identify the product. Write the equation for the reaction.

b Draw structural formulae for ethanol, ethanoic acid and ethyl ethanoate.

[L]

19 Analysis

Newspaper Tuesday 24th November

"A number of dark blue canisters have been found on local beaches. Police believe that they may contain highly toxic potassium cyanide. They are thought to have been washed from the decks of the Greek freighter *Alpha-Omega* during last weekends' violent storms. Anyone finding similar canisters is advised not to touch them, but to inform the police immediately".

Newspaper Friday 27th November

"Police scientists report that the canisters found on local beaches earlier this week did not contain potassium cyanide as feared. They did in fact contain coconut oil."

When stories like this appear in our newspapers, it means that somewhere a chemist has been busy finding out what was in those canisters. It is important that chemists know how to identify unknown substances. It is important that you know how to decide safely whether an unlabelled bottle in your laboratory contains distilled water or concentrated nitric acid. It is useful for you to know how to find out if some white crystals are sodium chloride or lead(II) nitrate. *Qualitative analysis is the name that chemists give to identifying unknown substances.* This chapter is concerned with the qualitative analysis of substances that you find in the laboratory. Most of these substances are ionic. You will need to carry out separate tests to identify the positive ions (cations) and negative ions (anions).

19.1 Starting analysis

When you are attempting to identify an unknown substance you will need to use much the same skills as a detective investigating a crime. You must be *organised* and *alert*. There are three main ways in which you will get information about an unknown substance:

1. A colour change may provide information. *Example:* If a substance is yellow when hot and white when cold it is likely to contain zinc ions (Zn^{2+}).

2. Forming a gas may provide information. *Example:* If sulphur dioxide gas (choking smell) is formed when hydrochloric acid is added, the unknown substance most likely contains sulphite ions (SO_3^{2-}).

3. Forming a precipitate may provide information. *Example:* If sodium hydroxide solution forms a brown jelly-like precipitate when added to unknown solution, that solution must contain n(III) ions (Fe^{3+}).

Be organised

Often when testing an unknown substance a gas is formed. It is of little use watching bubbles of gas escaping from a test tube and then thinking that you need to test the gas. You must be organised: you must be prepared! You must have all the chemicals you need for testing gases ready at the start of your analysis.

Table 1 shows some of the tests used to identify gases.

Name of gas	Tests for the gas
ammonia	1. Characteristic smell 2. Turns damp red litmus paper blue 3. Forms a thick white smoke with hydrogen chloride
carbon dioxide	Turns lime water milky
chlorine	1. Choking irritating smell 2. Turns damp blue litmus paper red then white
hydrogen	Burns, often with a pop
hydrogen chloride	1. Sharp stinging smell 2. Turns damp blue litmus paper red and forms a thick white smoke with ammonia
nitrogen dioxide	A brown gas that turns damp blue litmus paper red
oxygen	Relights a glowing splint
sulphur dioxide	1. Characteristic unpleasant smell 2. Turns orange potassium dichromate(VI) solution to a green colour
water vapour	1. Turns blue cobalt chloride paper pink 2. Turns anhydrous copper(II) sulphate from white to blue

Table 1

If a gas is formed when you are testing an unknown substance you should first *carefully* smell the gas. This may give you a clue as to which gas it is. However, **it is safe to assume that a gas with an unpleasant smell is a poisonous gas.** It is much better to identify a gas by using chemicals. Fig 19.1 shows some of the ways in which gases can be tested.

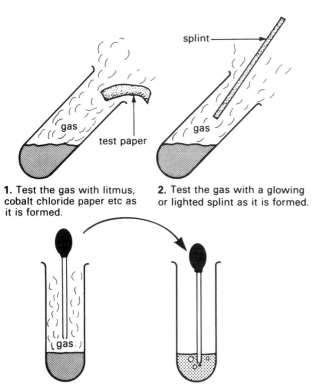

1. Test the gas with litmus, cobalt chloride paper etc as it is formed.

2. Test the gas with a glowing or lighted splint as it is formed.

3. Collect some of the gas in a teat pipette as it forms and pass it into a test solution, eg lime water.

Fig 19.1 Testing gases

Be alert

When you are given an unknown substance for the first time look at it carefully. Its appearance may give you a number of clues. Table 2 shows some of the deductions that you could make from the colour of an unknown substance.

Colour of substance	Deduction
grey	it may be a metal
blue	it may contain copper(II) ions (Cu^{2+})
green	it may contain copper(II) ions (Cu^{2+})
	it may contain iron(II) ions (Fe^{2+})
yellow	it might contain iron(III) ions (Fe^{3+})
white	it is most likely not a transition metal compound

Table 2

19.2 Testing for cations

1. *Reaction with sodium hydroxide*

The cations of many compounds are identified by testing with sodium hydroxide solution as shown in Fig 19.2.

In this test the precipitates formed are insoluble metal hydroxides. They are formed by metal ions from the unknown substance reacting with hydroxide ions from the sodium hydroxide solution:

Example:

$$\begin{array}{ccc} \text{copper(II)} + & \text{hydroxide} & \rightarrow \text{copper(II)} \\ \text{ions} & \text{ions} & \text{hydroxide} \end{array}$$

$$Cu^{2+}(aq) + 2OH^-(aq) \rightarrow Cu(OH)_2(s)$$

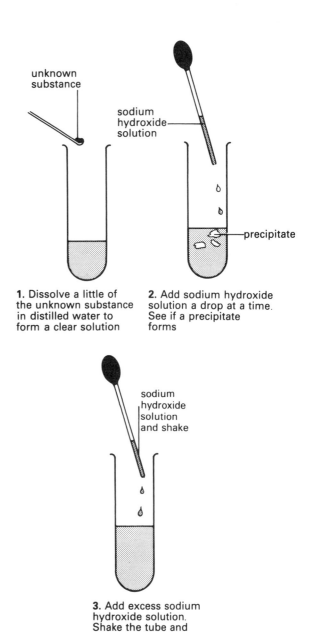

1. Dissolve a little of the unknown substance in distilled water to form a clear solution

2. Add sodium hydroxide solution a drop at a time. See if a precipitate forms

3. Add excess sodium hydroxide solution. Shake the tube and see if the precipitate dissolves in excess sodium hydroxide solution

Fig 19.2

Table 3 overleaf shows the results that can be expected if the unknown substance contains certain cations.

2. *Reaction with ammonia solution (aqueous ammonia)*

Solutions of ammonia in water can also be used to test for cations in solutions. The reactions are very similar to those with sodium hydroxide solution. Precipitates of the metal hydroxide are formed. However, there are some important differences. Table 4 overleaf shows the results that can be expected when certain cations are present.

You should realise by looking at Tables 3 and 4 that some cations like copper(II) or iron(III) can be identified using either sodium hydroxide solution or ammonia solution. However to identify zinc ions or lead ions both sodium hydroxide solution and ammonia solution must be used.

Cation present in unknown substance	Formula of cation	Observation on adding a few drops of sodium hydroxide solution	Observation on adding an excess of sodium hydroxide solution
aluminium	Al^{3+}	White ppt forms	Ppt dissolves to form a colourless solution
calcium	Ca^{2+}	Faint white ppt forms	No change
copper(II)	Cu^{2+}	Pale blue ppt forms	No change
iron(II)	Fe^{2+}	Dirty blue-green ppt forms	No change
iron(III)	Fe^{3+}	Brown ppt forms	No change
lead(II)	Pb^{2+}	White ppt forms	Ppt dissolves to form a colourless solution
zinc	Zn^{2+}	White ppt forms	Ppt dissolves to form a colourless solution

Table 3

Cation present in unknown substance	Formula of cation	Observation on adding a few drops of ammonia solution	Observation on adding an excess of ammonia solution
aluminium	Al^{3+}	White ppt forms	No change
calcium	Ca^{2+}	No ppt	No change
copper(II)	Cu^{2+}	Pale blue ppt forms	Ppt dissolves to form blue solution
iron(II)	Fe^{2+}	Dirty blue-green ppt forms	No change
iron(III)	Fe^{3+}	Brown ppt forms	No change
lead(II)	Pb^{2+}	White ppt forms	No change
zinc	Zn^{2+}	White ppt forms	Ppt dissolves to form a colourless solution

Table 4

Identifying the peculiar cations

Most cations are metal ions, but there are two exceptions. These are the hydrogen ion (H^+) and the ammonium ion (NH_4^+).

Testing for hydrogen ions

The hydrogen ion is the ion responsible for acidity. A solution must therefore contain hydrogen ions if:

1. It turns litmus paper red.
2. It turns Universal indicator red, orange or yellow.
3. It reacts with any carbonate producing carbon dioxide gas.

Testing for ammonium ions

Any compound containing ammonium ions will react with sodium hydroxide solution to form ammonia gas. This reaction is used as a test for ammonium ions. The unknown substance is warmed with sodium hydroxide solution and any gas formed is tested with damp red litmus paper.

19.3 Testing for anions

A lot of information about the anions present in an unknown substance can be obtained just by heating it.

Table 5 shows some of the results that can be expected if certain anions are present in the unknown substance.

1. *Heating the solid*

Name of anion	Formula of anion	Possible observations on heating
Carbonate	CO_3^{2-}	Carbon dioxide gas may be produced
Sulphite	SO_3^{2-}	Sulphur dioxide gas may be produced
Nitrate	NO_3^-	Oxygen gas may be produced *or* oxygen gas and nitrogen dioxide gas may be produced

Table 5

2. *Reaction with hydrochloric acid*

By adding dilute hydrochloric acid to an unknown substance you can find out if it contains *carbonate* or *sulphite* ions.

Carbonates produce carbon dioxide gas. This can be detected using lime water.

Sulphites produce sulphur dioxide gas. This can be detected using potassium dichromate(VI) solution. It may be necessary to warm the mixture to free the sulphur dioxide gas.

3. *Reaction with silver nitrate solution*

Silver nitrate solution is used to test *solutions* of unknown substances. It enables *chloride* ions to be identified, as shown in Fig 19.3.

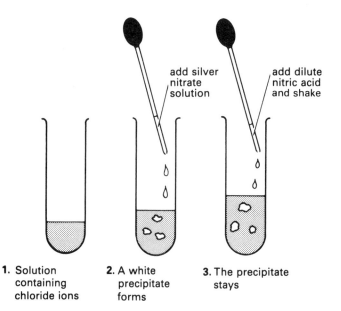

1. Solution containing chloride ions
2. A white precipitate forms
3. The precipitate stays

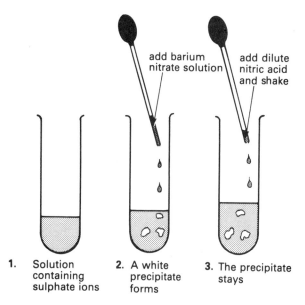

1. Solution containing sulphate ions
2. A white precipitate forms
3. The precipitate stays

Fig 19.3

The white precipitate of silver chloride will dissolve in concentrated ammonia solution.

Silver nitrate solution is also used to identify bromide ions (Br^-) and iodide ions (I^-) in aqueous solution.

If a solution contains bromide ions, a creamy yellow precipitate will form when silver nitrate solution is added. This precipitate will not dissolve in dilute nitric acid but some of the precipitate will dissolve in concentrated ammonia solution.

If a solution contains iodide ions, a yellow precipitate will form when silver nitrate solution is added. This precipitate will not dissolve in dilute nitric acid and it will not dissolve in concentrated ammonia solution.

The presence of bromide or iodide ions in solution is also confirmed by adding chlorine water, and then shaking the mixture with a small volume of 1,1,1-trichloroethane. Bromide ions are oxidised by chlorine to form bromine which gives a red-brown colour in the lower layer. Iodide ions are oxidised to iodine and a purple colour is seen in the lower layer.

4. *Reaction with barium nitrate solution or barium chloride solution*

Barium nitrate (or barium chloride), like silver nitrate, can be used to test for anions in *solutions*. Barium nitrate solution can be used to identify *sulphate* ions as shown in Fig 19.4.

5. *Reaction with Devarda's alloy and sodium hydroxide*

These reagents are used to test for the presence of *nitrate* ions in solution. A few cm³ of the solution to be tested is treated with an equal volume of dilute sodium hydroxide and then about 0·5 g of Devarda's alloy.

Fig 19.4

The mixture is then warmed gently until it boils. A piece of damp indicator paper placed *near* the mouth of the test tube is then used to test for ammonia.

Devarda's alloy together with sodium hydroxide reduces nitrate ions to ammonia.

19.4 Quantitative analysis

Sometimes chemists know what substances they are dealing with, but they need to find out *how much* of it there is. They need to use *quantitative* and not qualitative analysis.

Chemists often need to know the concentrations of acids and alkalis in aqueous solutions. It may be that the concentration of an acid bath for cleaning metal surfaces needs to be checked. It may be that the acidic waste from a factory has to be neutralised by alkali before it is allowed to run into rivers or into the sea. Chemists use pipettes and burettes to find out the concentration of acids or alkalis in solution. This technique is called *titration*.

Suppose the concentration of an alkaline solution needs to be found. The following stages are involved using the apparatus shown in Fig 19.7.

1. Make up a solution of an acid of known concentration (this is called a *standard solution*).
2. Fill a burette up to above the zero mark with this solution. Then open the tap and run out sufficient liquid so that it comes down to the zero mark.
(The bottom curve of the liquid should be level with the zero mark.)
3. Pipette 25 cm³ of the alkali solution into a conical flask.
4. Add two drops of a suitable indicator.
5. Run in acid from the burette until the alkali is just neutralised as shown by the indicator.
6. Note the volume of acid that has been added.

7. Repeat the experiment until accurate results are obtained.
8. Calculate the concentration of the alkali solution.

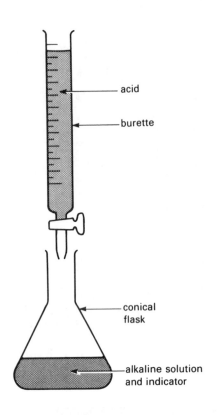
- acid
- burette
- conical flask
- alkaline solution and indicator

Fig 19.7

The following are typical calculations found in this type of work.

1. An aqueous solution of sulphuric acid contains $9.8\,g/dm^3$ of H_2SO_4. $20.0\,cm^3$ of this solution was needed to neutralise $25.0\,cm^3$ of a solution of potassium hydroxide in water. Calculate:

a the concentration, in mol/dm^3 of the sulphuric acid solution,
b the concentration, in mol/dm^3 of the potassium hydroxide solution,
c the concentration, in g/dm^3 of the potassium hydroxide solution.

1 mole of $H_2SO_4 = 2 \times 1 + 32 + 4 \times 16 = 98\,g$

$$9.8\,g \text{ of } H_2SO_4 \;=\; \frac{9.8}{98} = 0.1 \text{ mole}$$

Therefore the concentration of the sulphuric acid is $0.1\,mol/dm^3$. The equation for the reaction is:

$$H_2SO_4 + 2KOH \longrightarrow K_2SO_4 + 2H_2O$$
1 mole   2 moles   1 mole   2 moles

$1000\,cm^3$ of sulphuric acid solution contains 0.1 mole
$1\,cm^3$ of sulphuric acid solution contains

$$\frac{0.1}{1000}\,mol$$

$20\,cm^3$ of sulphuric acid solution contains

$$\frac{0.1}{1000} \times 20 \text{ mol } = 0.002 \text{ mol}$$

Therefore 0.002 mol of sulphuric acid are used in the titration. Since 1 mole of sulphuric acid reacts with 2 moles of potassium hydroxide (see equation).

$0.002 \times 2 = 0.004$ mol of potassium hydroxide must be used in the titration.

$25\,cm^3$ of the alkali solution contain 0.004 mol of KOH
$1\,cm^3$ of the alkali solution contain

$$\frac{0.004}{25}\,mol \text{ of KOH}$$

$1000\,cm^3$ of the alkali solution contain

$$\frac{0.004}{25} \times 1000 \text{ mol of KOH} = 0.16 \text{ mol of KOH}$$

Therefore the concentration of the potassium hydroxide solution is $0.16\,mol/dm^3$
1 mole of KOH $= 39 + 16 + 1 = 56\,g$
Therefore 0.16 mol of KOH $= 0.16 \times 56 = 8.96\,g$

Therefore the concentration of the potassium hydroxide solution is $0.16\,mol/dm^3$

2. You are provided with a solution of hydrochloric acid containing $3.65\,g/dm^3$ of HCl. $22.4\,cm^3$ of this solution is needed to neutralise $25.0\,cm^3$ of aqueous sodium hydroxide. Calculate:

a the concentration, in mol/dm^3 of the hydrochloric acid,
b the concentration, in mol/dm^3 of the sodium hydroxide solution,
c the concentration, in g/dm^3 of the sodium hydroxide solution.

1 mole of HCl $= 1 + 35.5 = 36.5\,g$

$$3.65\,g \text{ of HCl } = \frac{3.65}{36.5} = 0.10 \text{ mol}$$

Therefore the concentration of the hydrochloric acid $= 0.1\,mol/dm^3$

The equation for the reaction is:

HCl $+$ NaOH $\longrightarrow$ NaCl $+$ H$_2$O
1 mole   1 mole   1 mole   1 mole

$1000\,cm^3$ of the hydrochloric acid solution contains 0.1 mol of HCl
$1\,cm^3$ of the hydrochloric acid solution contains

$$\frac{0.1}{1000}\,mol \text{ of HCl}$$

$22.4\,cm^3$ of the hydrochloric acid solution contains

$$\frac{0.1}{1000} \times 22.4 \text{ mol of HCl} = 0.00224 \text{ mol}$$

Therefore 0.00224 mol of HCl are used in the titration. Since 1 mole of HCl reacts with 1 mole of NaOH (see equation), 0.00224 moles of NaOH must be used in the titration.
$25\,cm^3$ of the alkali solution contains 0.00224 moles of NaOH
$1\,cm^3$ of the alkali solution contains

$$\frac{0.00224}{25}\,mol \text{ of NaOH}$$

1000 cm³ of the alkali solution contains

$\frac{0.00224}{25} \times 1000$ moles of NaOH

= 0·090 moles (to 3 significant figures)

The concentration of the sodium hydroxide solution is therefore 0·090 mol/dm³.

1 mole of NaOH = 23 + 16 + 1 = 40·0 g.

Therefore 0·090 moles of NaOH = 0·090 × 40 = 3·60 g.

Therefore the concentration of the sodium hydroxide solution is 3·60 g/dm³.

Questions

1. Aqueous ammonia was added to a colourless solution and a white precipitate was formed. The precipitate dissolved in excess aqueous ammonia. The experiment was repeated using aqueous sodium hydroxide in place of aqueous ammonia. Again, a white precipitate was formed which was soluble in excess aqueous sodium hydroxide. Which one of the following ions was present in the solution?

A Al^{3+}
B Ca^{2+}
C Pb^{2+}
D Mg^{2+}
E Zn^{2+}

2. When a white solid X was heated, a brown gas was given off and the solid turned brown. The residue on cooling was yellow. Which one of the following could X be?

A copper(II) nitrate
B iron(III) nitrate
C lead(II) nitrate
D potassium nitrate
E zinc nitrate

3. Which one of the following tests would you use to distinguish between sodium sulphate solution and sodium carbonate solution?

A add barium chloride solution
B add dilute hydrochloric acid
C add lead(II) nitrate solution
D add silver nitrate solution
E add sodium hydroxide solution

4. What is the concentration, in mol/dm³, of a solution containing 1 g of sodium hydroxide in 100 cm³ of solution?

A 0·05 mol/dm³
B 0·01 mol/dm³
C 0·15 mol/dm³
D 0·20 mol/dm³
E 0·25 mol/dm³

5. 20 cm³ of 0·15 mol/dm³ sodium hydroxide solution exactly neutralised 10 cm³ of 0·10 mol/dm³ of an organic acid solution, $H_8C_6O_7$. What is the basicity of this organic acid?

A 1
B 2
C 3
D 4
E 5

6. AMMONIA, CARBON DIOXIDE, CHLORINE, HYDROGEN, HYDROGEN CHLORIDE, OXYGEN, SULPHUR DIOXIDE.

From the above list of gases choose one which:

a is less dense than air
b turns acidified potassium dichromate(VI) solution green
c relights a glowing splint
d burns in air
e dissolves in water to form an acidic solution
f dissolves in water to form an alkaline solution
g turns lime water milky
h fumes in damp air
i bleaches damp litmus paper
j is formed when sulphur burns.

7. Name:

a two metal carbonates which are insoluble in water
b a metal chloride which is insoluble in water
c a metal sulphate which is insoluble in water
d a metal hydroxide which is soluble in water
e a red brown metal hydroxide
f a metal which colours a bunsen flame red
g a blue metal hydroxide
h a metal nitrate which forms brown fumes on heating
i a metal nitrate which does not form brown fumes on heating.

8. You have three unlabelled bottles. One contains sodium chloride, one contains sodium carbonate and the third contains a mixture of sodium chloride and sodium carbonate. How would you find out which is which?

9. Suppose you are given a solution of copper(II) chloride in water.

a What colour would you expect the solution to be?
b Describe what you would expect to observe when the following are added to separate samples of the copper(II) chloride solution:
   (i) silver nitrate solution
   (ii) sodium hydroxide solution
   (iii) dilute hydrochloric acid
   (iv) ammonia solution.

10.

a For each of the statements below, name *one* substance which behaves as described.

A is a colourless gas that reacts vigorously with chlorine when exposed to bright light.
B is a silver-grey metal that reacts very slowly with cold water but rapidly when heated in steam.
C is a white crystalline solid that, on heating, gives off steam followed by two other gases one of which rekindles a glowing splint and the other is brown.
D is a colourless liquid that does not mix with water and that dissolves paraffin wax.
E is a red-brown powder that is unaffected by hydrochloric acid but dissolves in dilute nitric acid, giving off a gas which turns brown in contact with air.
F is a white solid that reacts vigorously with water to produce a strongly acidic solution.

Give equations for *any two* of the above reactions.

b Crystalline sodium carbonate has the formula:

$$Na_2CO_3.xH_2O$$

When 4·29 g of the crystals are heated to constant mass the residue has a mass of 1·59 g. Calculate the value of x and explain briefly how "heating to constant mass" is carried out. (Relative atomic masses: H, 1·0; C, 12; O, 16; Na, 23.)

[C]

11. In two titration experiments it was found that:

I 20 cm³ of 0·1 mol/dm³ potassium chromate(VI) (K_2CrO_4) required 20 cm³ of 0·2 mol/dm³ silver nitrate for complete reaction,

II 20 cm³ of 0·1 mol/dm³ potassium chromate(VI) required 10 cm³ of 0·2 mol/dm³ barium chloride for complete reaction.

Silver chromate(VI) and barium chromate(VI) are both insoluble in water.

a How many moles of silver nitrate are required to react completely with one mole of potassium chromate(VI)?

b How many moles of barium chloride are required to react completely with one mole of potassium chromate(VI)?

c Write equations for the reactions taking place in Experiments I and II.

d Calculate the mass of silver chromate(VI) precipitated in Experiment I and the mass of barium chromate(VI) precipitated in Experiment II.

e Write the equation for the reaction between solutions of barium chloride and silver nitrate and calculate the volume of 0·2 mol/dm³ silver nitrate required to react completely with 20 cm³ of 0·2 mol/dm³ barium chloride.

f Give the formulae of three halides of silver that are insoluble in water.

[C]

12. Imagine that you are provided with solid samples of calcium carbonate, hydrated copper(II) sulphate, anhydrous iron(III) chloride and potassium sulphite.

a Choose any *two* of these solid samples and show how you would confirm, by chemical tests, the presence of both ions in each. (This means that you have to identify four ions in all.)

b Choose any one of the four solids and explain how you would obtain a sample of metal from it, giving equations where appropriate.

c Calcium carbonate and hydrated copper(II) sulphate are decomposed by heat. What would you expect to *see* when each is heated? How would you identify any gas which is produced?
Write the equation for the reaction in each case.

[C]

13.

a An excess of sodium hydrogencarbonate was shaken with about 150 cm³ of water in a stoppered flask, at room temperature to produce a saturated solution. The contents of the flask were filtered into a clean, dry flask, using a dry funnel fitted with a dry filter paper.

To determine the volume of a solution of hydrochloric acid containing 10·95 g/dm³ needed to neutralise 10·0 cm³ of the sodium hydrogencarbonate solution, four titrations were carried out. In each case the first burette reading was 0·0 cm³ and the level of acid in the burette at the end-point was as shown opposite.

(i) From the burette readings, what is the volume of acid needed to neutralise 10·0 cm³ of the sodium hydrogencarbonate solution?

(ii) Calculate, from the information given, the concentration of the acid in mol/dm³.

(iii) The equation for the reaction is:

$$NaHCO_3 + HCl \rightarrow$$
$$NaCl + H_2O + CO_2.$$

Hence calculate from the answer in (i) the concentration of the sodium hydrogencarbonate in mol/dm³.

(iv) Calculate the concentration of the sodium hydrogencarbonate in g/dm³.

b (i) Why is it necessary to filter the solution in **a** through a dry filter paper into a dry flask?

(ii) Why must the solution be prepared at room temperature?

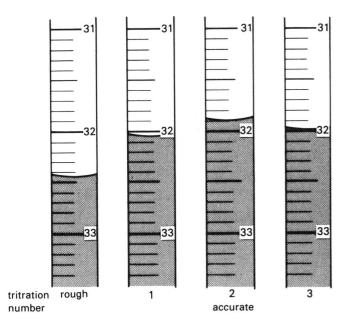

tritration number — rough — 1 — 2 — 3
accurate

c Addition of a few drops of the saturated solution of sodium hydrogencarbonate to a few cm³ of aqueous calcium chloride caused no visible change, but on warming the mixture a white precipitate appeared. Explain these observations.

(Relative atomic masses: H, 1·0; C, 12·0; O, 16·0; Cl, 35·5; Na, 23·0.)

[C]

14. In each of sections **a, b** and **c** below, give the name of *one* possible substance indicated by each letter and write equations for *two* of the reactions.

a When a green solid A was heated, a black solid and a gas B were formed. When gas B was passed into aqueous calcium hydroxide, a white precipitate C was obtained.

b When a white solid P was heated, brown fumes Q were given off and a yellow solid was left which turned white on cooling. When the brown fumes were passed into water, a colourless solution R was formed.

c X is a white solid which is soluble in water. When aqueous sodium hydroxide was added to a solution of X, a white precipitate Y was formed which redissolved when more of the alkali was added forming a solution Z. On adding dilute nitric acid and silver nitrate solution to a solution of X, another white precipitate was formed.

[JMB]

Data page

Element	Symbol	Atomic number	Relative atomic mass	Density at 20 °C (g/cm³)	Good or bad conductor of electricity	Melting point /°C	Boiling point /°C
Aluminium	Al	13	27	2·7	Good	660	2470
Argon	Ar	18	40	1·7 g/dm³	Bad	−189	−186
Arsenic	As	33	75	5·7	Good	sublimes at 610	
Barium	Ba	56	137	3·5	Good	710	1640
Bromine	Br	35	80	3·1	Bad	−7	58
Calcium	Ca	20	40	1·5	Good	850	1487
Carbon	C	6	12	2·3	Good (graphite) Bad (diamond)	3500	3900
Chlorine	Cl	17	35·5	1·5 g/dm³	Bad	−101	−34
Chromium	Cr	24	52	7·1	Good	1900	2482
Cobalt	Co	27	59	8·9	Good	1492	2900
Copper	Cu	29	64	9·0	Good	1083	2580
Fluorine	F	9	19	0·8 g/dm³	Bad	−220	−188
Gallium	Ga	31	70	6·0	Good	30	2400
Germanium	Ge	32	73	5·3	Good	937	2830
Gold	Au	79	197	19·3	Good	1063	2970
Helium	He	2	4	0·2 g/dm³	Bad	−270	−269
Hydrogen	H	1	1	0·1 g/dm³	Bad	−259	−253
Iodine	I	53	127	4·9	Bad	114	183
Iron	Fe	26	56	7·9	Good	1539	3000
Krypton	Kr	36	84	3·5 g/dm³	Bad	−157	−153
Lead	Pb	82	207	11·3	Good	327	1750
Lithium	Li	3	7	0·5	Good	180	1330
Magnesium	Mg	12	24	1·7	Good	650	1100
Manganese	Mn	25	55	7·4	Good	1250	2100
Mercury	Hg	80	201	13·6	Good	−39	357
Neon	Ne	10	20	0·8 g/dm³	Bad	−249	−246
Nickel	Ni	28	59	8·9	Good	1453	2730
Nitrogen	N	7	14	1·2 g/dm³	Bad	−210	−196
Oxygen	O	8	16	1·3 g/dm³	Bad	−219	−183
Phosphorus	P	15	31	1·8	Bad	44	280
Platinum	Pt	78	195	21·4	Good	1769	3800
Potassium	K	19	39	0·9	Good	63	774
Scandium	Sc	21	45	3·0	Good	1540	2730
Selenium	Se	34	79	4·8	Good	217	685
Silicon	Si	14	28	2·4	Good	1410	2360
Silver	Ag	47	108	10·5	Good	961	2210
Sodium	Na	11	23	1·0	Good	98	883
Strontium	Sr	38	88	2·6	Good	770	1380
Sulphur	S	16	32	2·1	Bad	119	445
Tin	Sn	50	119	7·3	Good	232	2270
Titanium	Ti	22	48	4·5	Good	1680	3260
Vanadium	V	23	51	6·1	Good	1920	3000
Uranium	U	92	238	19·0	Good	1133	3800
Zinc	Zn	30	65	7·1	Good	419	907

Index